Mathematics

Essential Skills

5

OXFORD
UNIVERSITY PRESS

253 Normanby Road, South Melbourne, Victoria 3205, Australia

Oxford University Press is a department of the University of Oxford.
It furthers the University's objective of excellence in research, scholarship, and education by publishing worldwide in

Oxford New York

Auckland Cape Town Dar es Salaam Hong Kong Karachi
Kuala Lumpur Madrid Melbourne Mexico City Nairobi
New Delhi Shanghai Taipei Toronto

With offices in

Argentina Austria Brazil Chile Czech Republic France Greece Guatemala
Hungary Italy Japan Poland Portugal Singapore South Korea Switzerland
Thailand Turkey Ukraine Vietnam

First published 2010
Reprinted 2014, 2018(D)

ISBN 978 0 19 557390 9

Typeset by Palmer Higgs
Printed and bound in Australia by Ligare Book Printers Pty Ltd

PAPUA NEW GUINEA

Mathematics

Essential Skills

5

Pat Lilburn

OXFORD
UNIVERSITY PRESS
AUSTRALIA & NEW ZEALAND

Contents

For Students 1

Number and Application
Skills practice 2

Assessment 42

Measurement
Skills practice 49

Assessment 65

Space and Shape
Skills practice 67

Assessment 76

Chance and Data
Skills practice 78

Patterns
Skills practice 80

Assessment (for Chance and Data/Patterns) 83

Important Facts 84

Answers 86

For Students

The *Oxford Essential Skills Book* has been written to support the *Oxford Grade 5 Mathematics Student Books A* and *B*.

The focus of this book is on providing practice examples to revise and support the concepts learned in the two Student Books.

The book is set out under the five Strands outlined in the *Lower Primary Mathematics Syllabus* and *Teacher Guide*:

Number and Application
Measurement
Space and Shape
Chance and Data
Patterns

Within each Strand, you will find many practice examples that relate directly to the Learning Outcomes for that Strand. These Learning Outcomes are specified in the *Lower Primary Mathematics Teacher Guide*.

The subheading in each Strand shows you what the examples will help you practise.

5.1.1 Order, read and write four and five-digit numbers

The Help Box contains worked examples to show you how to set out and complete the examples in each Strand. The Remember Box contains information that will help you complete the examples.

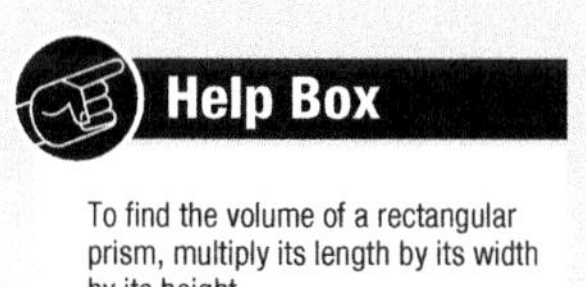

To find the volume of a rectangular prism, multiply its length by its width by its height.

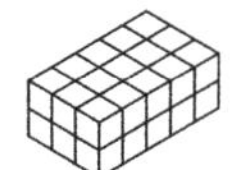

Length = 5 cm
Width = 3 cm
Height = 2 cm

Volume = Length × Width × Height
$= 5 \times 3 \times 2 = 30\ cm^3$

The > sign means 'more than' and the < sign means 'less than'.

At the end of each Strand there is an Assessment section that covers all the examples from that Strand. If you have difficulty with any test questions, go back to the relevant page in the Strand and look at the worked examples before doing the test again.

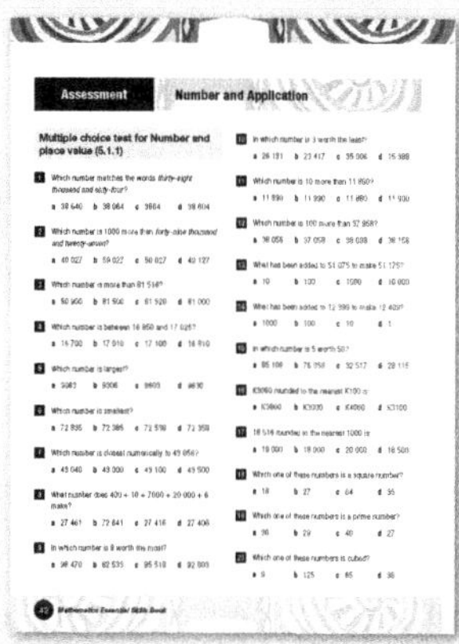

Assessment Number and Application

Multiple choice test for Number and place value (5.1.1)

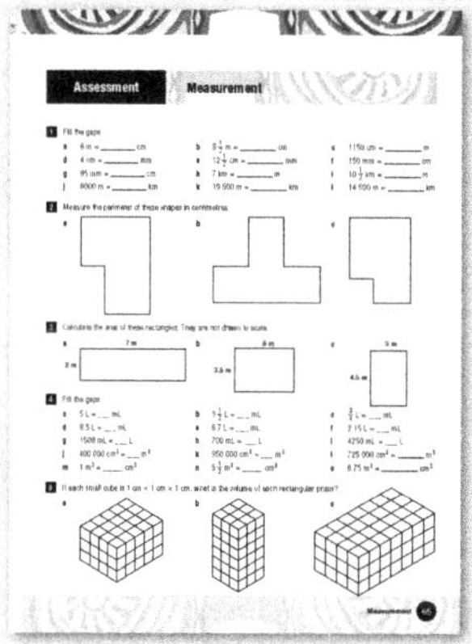

Assessment Measurement

There is an Answers section and a section for Important Facts at the back of the book.
The Answers section contains answers to all practice examples and test questions while the Important Facts section contains some of the facts that you will need to refer to throughout the year.

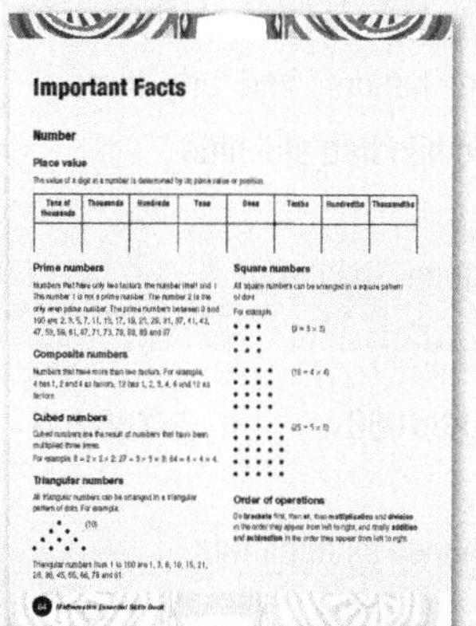

Important Facts

Number

Place value

Prime numbers

Square numbers

Composite numbers

Cubed numbers

Triangular numbers

Order of operations

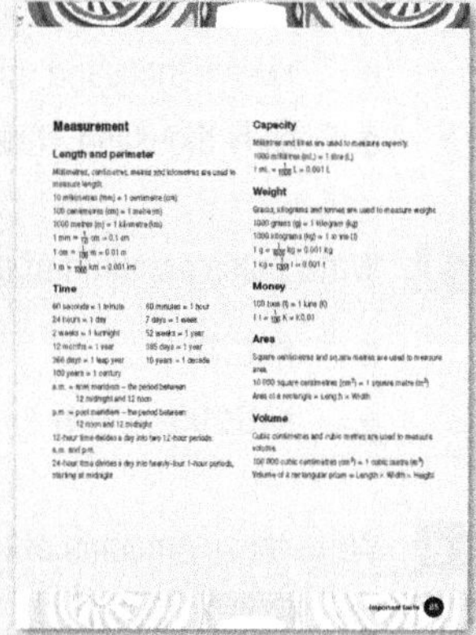

Measurement

Length and perimeter

Time

Capacity

Weight

Money

Area

Volume

Strand Number and Application

5.1.1 Order, read and write four and five-digit numbers

Read and write four and five-digit numbers in words and numerals

1 Write these words as numerals.

- **a** seven thousand, four hundred and thirty-six
- **b** two thousand, seven hundred and fifty-eight
- **c** five thousand, nine hundred and seventy-one
- **d** nine thousand, three hundred and twenty
- **e** three thousand, eight hundred and ninety-seven
- **f** fifteen thousand, six hundred and eighty-two
- **g** forty-six thousand and sixty-five
- **h** twenty-four thousand, one hundred and seven
- **i** seventy thousand, five hundred and forty-nine
- **j** sixty-eight thousand, two hundred and ninety-three
- **k** thirty-seven thousand, three hundred and fifty
- **l** fifty-nine thousand and five

2 Write these numerals as words.

a	6783	**b**	3081	**c**	9775	**d**	2150	**e**	5442	**f**	1209
g	16 552	**h**	81 340	**i**	33 706	**j**	50 027	**k**	28 643	**l**	62 114

3 Write in words the number that is 1000 more than these numbers.

a	8406	**b**	1550	**c**	7963	**d**	4818	**e**	2497	**f**	9104
g	42 317	**h**	11 056	**i**	89 745	**j**	25 608	**k**	19 560	**l**	70 031

4 Write in numerals the number that is 1000 less than these numbers.

- **a** seven thousand, eight hundred and ninety-four
- **b** three thousand and sixty-seven
- **c** twelve thousand, nine hundred and forty-three
- **d** thirty-six thousand, one hundred and seventy-five
- **e** sixty thousand, three hundred and nine
- **f** forty-three thousand, two hundred and seventeen

5 Write these amounts of money in words.

a	K5670	**b**	K3096	**c**	K15 985	**d**	K63 147	**e**	K49 678	**f**	K54 032
g	K19 299	**h**	K30 080	**i**	K85 200	**j**	K27 248	**k**	K73 155	**l**	K91 218

6 Write these amounts of money as numerals.

- **a** Twenty-five thousand, two hundred and ninety kina
- **b** Seventy thousand, four hundred and sixty-five kina
- **c** Eighteen thousand, seven hundred and twenty-five kina
- **d** Fifty-eight thousand and forty-three kina

Compare and order four and five-digit numbers

1 Write the numbers that are more than 9875.

8750 9890 11 265 9748 9950
10 063 6942 9880 19 218 9000
9800 9910 9871 9900 10 125

2 Write the numbers that are less than 15 120.

15 000 10 175 15 510 12 758 20 740
15 110 15 250 15 010 15 115 15 150
15 125 14 999 15 099 15 210 10 512

3 Write the prices that are more than K78 056.

K78 006 K75 839 K78 040 K97 115 K79 100
K81 000 K75 016 K97 182 K69 599 K96 765
K60 658 K46 753 K89 506 K36 014 K78 060

4 Write the prices that are less than K35 550.

K36 199 K19 995 K20 100 K41 995 K35 510
K39 999 K35 560 K34 800 K35 600 K35 549
K35 555 K35 050 K33 550 K35 999 K30 500

5 Write the numbers that are between 9579 and 12 086.

9 500 11 755 12 800 9599 12 080
10 999 9810 9625 12 180 9550
12 010 12 900 12 095 9758 11 006

6 Write the numbers that are between 19 780 and 23 085.

20 600 23 079 19 726 19 544 19 785
22 162 23 099 18 554 19 350 21 738
19 363 19 677 19 792 19 070 23 016

7 Write a number that comes between each pair of numbers.

a	8016 ______ 9005		**b**	1598 ______ 2000		**c**	9562 ______ 10 008	
d	9999 ______ 11 000		**e**	87 950 ______ 87 990		**f**	58 000 ______ 58 155	
g	95 672 ______ 96 000		**h**	18 755 ______ 18 850		**i**	65 999 ______ 66 100	
j	75 000 ______ 75 050		**k**	89 000 ______ 89 155		**l**	36 999 ______ 37 010	

8 Write each set of numbers in order from smallest to largest.

a 5645, 6445, 5465, 6554, 4556
b 3018, 3180, 3080, 3810, 3081
c 8430, 8043, 8340, 8403, 8034
d 7912, 9721, 7129, 7219, 9127
e 12 373, 12 705, 12 037, 12 370, 12 703
f 31 688, 31 866, 31 060, 31 680, 31 806
g 23 912, 23 129, 23 291, 23 192, 23 219
h 59 007, 59 070, 59 017, 59 710, 59 010

9 Write three numbers in order that come between each pair of numbers.

a 7800, ______, ______, ______, 8095
b 4975, ______, ______, ______, 5060
c 39 890, ______, ______, ______, 41 120
d 78 016, ______, ______, ______, 79 005
e 81 598, ______, ______, ______, 82 000
f 69 562, ______, ______, ______, 71 008
g 87 950, ______, ______, ______, 87 990
h 58 000, ______, ______, ______, 58 155

10 Which of the two numbers on the right is closer numerically to the number on the left?

a	86 050	85 500	87 000	**b**	97 898	95 000	98 000
c	57 566	55 362	55 758	**d**	81 975	82 000	81 500
e	78 815	80 167	80 003	**f**	37 502	37 009	37 750
g	49 627	50 696	45 474	**h**	92 243	91 565	92 506
i	63 008	62 715	64 001	**j**	70 138	68 796	73 787

Remember

The > sign means 'more than' and the < sign means 'less than'.

11 Use > or < to fill the gap.

a	95 058	☐	90 508	**b**	45 776	☐	45 780
c	67 529	☐	67 520	**d**	30 010	☐	30 005
e	74 998	☐	74 909	**f**	59 065	☐	59 108
g	60 562	☐	61 500	**h**	27 660	☐	27 606
i	35 000	☐	34 958	**j**	15 900	☐	10 590
k	89 098	☐	89 100	**l**	43 260	☐	43 026
m	72 365	☐	72 315	**n**	31 780	☐	31 800
o	26 019	☐	26 020	**p**	54 169	☐	54 175

Place value with four and five-digit numbers

1 What number will these make?

a 600 + 70 + 2000 + 8
b 80 + 4000 + 2 + 500
c 9000 + 1 + 20 + 300
d 700 + 6 + 1000
e 7000 + 40 + 10 000 + 900 + 8
f 50 + 80 000 + 3 + 1000 + 200
g 500 + 60 000 + 70 + 2000
h 7 + 30 000 + 6000 + 400
i 5 + 900 + 20 000 + 80 + 3000
j 50 000 + 6 + 10 + 500 + 7000

2 Which numbers are these?

a 5 hundreds, 2 thousands, 7 ones and 4 tens
b 7 tens, 8 thousands, 3 ones and 1 hundred
c 6 ones, 4 thousands, 2 ten thousands, 1 ten and 3 hundreds
d 9 thousands, 2 hundreds, 8 ones, 6 ten thousands and 5 tens
e 1 thousand, 7 hundreds, 9 ten thousands and 2 ones
f 4 tens, 8 ten thousands, 5 ones, 3 hundreds and 5 thousands
g 7 ten thousands, 4 ones, 6 tens and 3 thousands
h 9 ones, 1 hundred, 2 thousands, 6 ten thousands and 4 tens
i 8 tens, 9 hundreds, 1 ten thousand and 1 thousand
j 8 thousands, 3 ones, 5 hundreds, 2 ten thousands and 9 tens

3 Write the number from each group in which 4 is worth the most.

a	5467	4500	6794	8541	9574
b	1409	2346	7408	4512	1274
c	94 616	57 428	36 041	45 076	14 679
d	36 042	49 666	74 150	23 064	37 418
e	99 463	64 512	83 041	49 828	87 416
f	74 869	30 427	40 339	60 064	36 415

4 Write the number from each group in which 9 is worth the least.

a	7890	9645	3900	1209	4912
b	1932	8769	4792	9044	3976
c	68 915	49 104	33 229	95 418	66 932
d	19 536	16 925	96 000	15 093	94 168
e	75 962	18 948	92 016	57 192	38 940
f	17 639	98 072	29 641	17 896	95 017

5 Write three numbers where:

a 5 is worth 5000
b 2 is worth 20
c 7 is worth 700
d 6 is worth 60
e 1 is worth 10 000
f 4 is worth 4
g 9 is worth 900
h 8 is worth 8000
i 3 is worth 30 000

6 Write the number that is 10 more than the number on the left.

a	5990	5999	5980	6000	5100	5991
b	12 395	12 400	12 405	12 390	13 000	12 495
c	20 997	21 007	21 097	20 007	20 107	21 997

7 Write the number that is 100 more than the number on the left.

a	3750	3760	3800	4750	3850	3751
b	49 963	49 973	50 063	49 063	50 963	49 163
c	15 008	15 108	15 018	16 008	15 009	16 010

8 Write if 10, 100 or 1000 has been added to each number to make the new number.

a 7305 to 7405
b 8796 to 8806
c 5912 to 6012
d 11 403 to 12 403
e 39 046 to 40 046
f 25 805 to 25 815
g 15 000 to 15 010
h 46 910 to 47 010
i 29 565 to 30 565
j 32 063 to 32 163
k 40 159 to 40 169
l 19 030 to 20 030
m 21 399 to 21 409
n 17 999 to 18 099
o 36 900 to 37 000

9 In which of these numbers is:

a 6 worth 6000?
b 6 worth 60 000?
c 2 worth 200?
d 2 worth 20?
e 2 worth 2?
f 6 worth 60?
g 6 worth 600?
h 2 worth 2000?

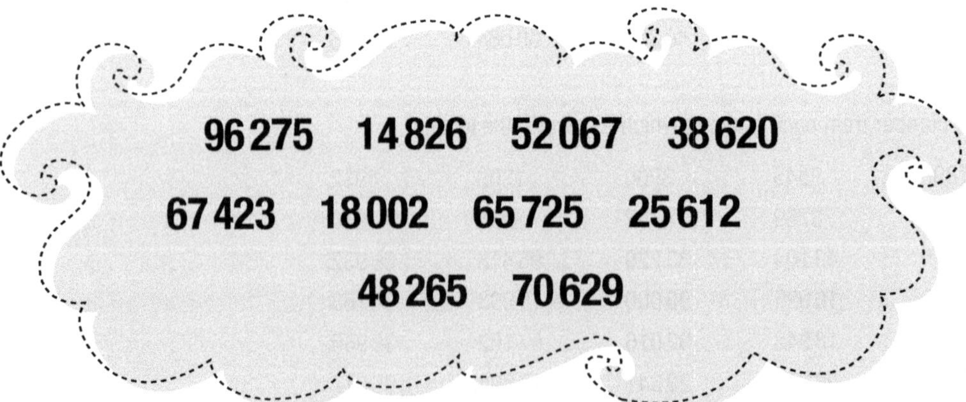

10 Write the number that is:

a	100 more than 5718	**b**	10 less than 9216	**c**	1000 less than 7530
d	10 more than 28 095	**e**	100 less than 94 529	**f**	1000 more than 31 233
g	100 more than 43 958	**h**	1 less than 78 000	**i**	100 more than 17 036
j	1000 less than 60 725	**k**	1 more than 87 039	**l**	10 less than 54 827
m	10 more than 64 396	**n**	100 less than 23 425	**o**	1 less than 19 672
p	1000 more than 49 136	**q**	10 less than 76 588	**r**	1000 less than 50 006
s	1 more than 59 999	**t**	10 more than 85 053	**u**	100 more than 93 647
v	10 less than 44 165	**w**	1000 more than 25 510	**x**	10 000 less than 36 004

11 Rearrange these five digits to make:

- **a** the largest number
- **b** the smallest number
- **c** the number closest to 30 000
- **d** the number closest to 55 000

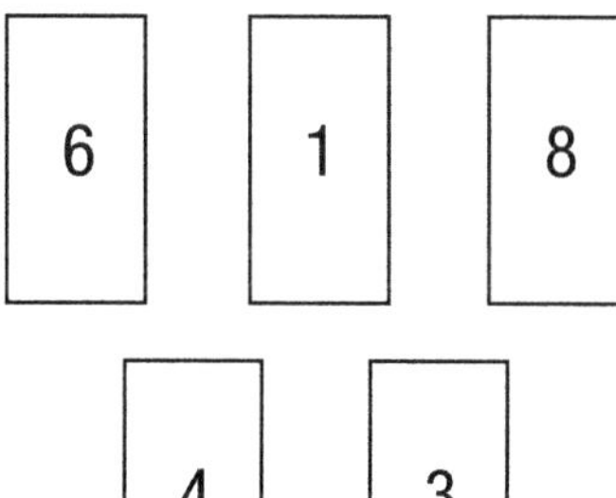

12 Rearrange these five digits to make:

- **a** the largest number
- **b** the smallest number
- **c** the number closest to 70 000
- **d** the number closest to 45 000

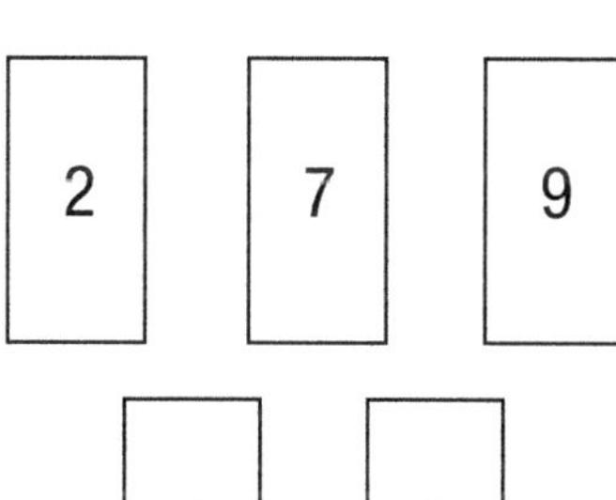

Round off whole numbers

1 Round these money amounts to the nearest K10.

a	K89	**b**	K62	**c**	K78	**d**	K36	**e**	K51	**f**	K45
g	K127	**h**	K715	**i**	K322	**j**	K559	**k**	K106	**l**	K288
m	K412	**n**	K531	**o**	K865	**p**	K157	**q**	K294	**r**	K376
s	K2546	**t**	K1344	**u**	K5675	**v**	K3081	**w**	K1896	**x**	K4109

2 Round these money amounts to the nearest K100.

a	K187	**b**	K232	**c**	K582	**d**	K395	**e**	K115	**f**	K271
g	K859	**h**	K425	**i**	K640	**j**	K773	**k**	K538	**l**	K710
m	K4352	**n**	K1825	**o**	K2661	**p**	K5080	**q**	K1206	**r**	K3015
s	K5924	**t**	K3890	**u**	K7150	**v**	K4678	**w**	K1109	**x**	K2555

3 Round these numbers to the nearest 1000.

a	4338	**b**	5752	**c**	7613	**d**	1955	**e**	2450	**f**	6005
g	57 587	**h**	81 306	**i**	38 874	**j**	61 755	**k**	48 963	**l**	11 216
m	26 439	**n**	79 820	**o**	34 500	**p**	17 056	**q**	27 595	**r**	53 714
s	72 866	**t**	60 183	**u**	42 561	**v**	14 619	**w**	67 540	**x**	31 508

4 Round these numbers to the nearest 10 000.

a	87 594	**b**	52 125	**c**	65 208	**d**	34 567	**e**	78 336	**f**	24 080
g	13 728	**h**	75 612	**i**	53 576	**j**	18 065	**k**	87 963	**l**	36 415
m	43 208	**n**	12 044	**o**	27 629	**p**	74 392	**q**	67 280	**r**	25 607
s	89 960	**t**	51 658	**u**	58 016	**v**	13 788	**w**	92 599	**x**	69 900

5 Which of these numbers could be rounded to 7000?

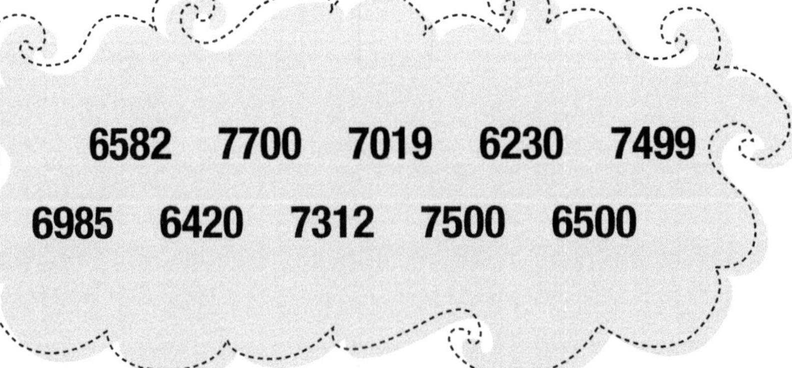

6 Which of these numbers could be rounded to 30 000?

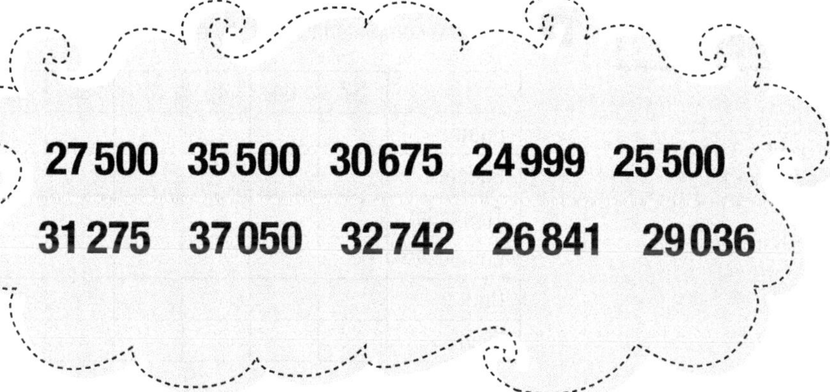

7 Copy and complete this chart by rounding each number as shown.

	Number	Nearest 10	Nearest 100	Nearest 1000	Nearest 10 000
a	14 684				
b	53 091				
c	38 712				
d	22 537				
e	85 175				
f	68 829				
g	12 418				
h	75 309				
i	47 685				
j	91 763				

Square, cubed, prime, composite, triangular, odd and even numbers

Remember

A composite number has factors other than itself and 1.
Example: factors of 8 are 1, 2, 4 and 8.

A prime number only has itself and 1 as factors.
Example: factors of 11 are 1 and 11.

1 Copy and complete the chart by ticking the spaces that fit each number.

	37	64	45	17	36	10	25	31	26
Prime									
Square									
Triangular									
Composite									
Odd									
Even									

Remember

A cubed number is the answer to a number multiplied three times.

Example: 2 cubed = 2 × 2 × 2 = 8 so 8 is a cubed number.

2 Write the square numbers from this group of numbers.

64 40 9 25 56 16 100 72 81 36 23 39

3 Write the cubed numbers from this group of numbers.

27 10 15 125 100 1000 9 64 16 36 343 24

4 Write the odd numbers from this group of numbers.

5607 4322 6980 4416 3021 7555 1978 8743 3117 9778 3159
14 325 56 713 28 976 63 024 32 270 51 617 75 582 88 995

5 Without calculating, write the number sentences that will give an even answer.

a 315 + 760 **b** 786 – 414 **c** 744 × 3 **d** 6135 – 3872
e 523 × 6 **f** 7652 + 8124 **g** 8645 × 7 **h** 5327 + 1083
i 17 688 – 18 348 **j** 1099 × 9 **k** 34 568 + 12 092 **l** 22 717 – 13 542
m 2448 × 4 **n** 45 481 + 67 373 **o** 72 167 – 36 928 **p** 5978 × 5

6 Without calculating, write the number sentences that will give an odd answer.

a 68 142 + 5469 + 16 555 **b** 3775 × 25 **c** 63 201 – 28 567
d 36 783 × 12 **e** 47 009 + 32 165 + 50 977 **f** 6883 × 33
g 6517 + 27 885 + 14 353 **h** 45 729 × 15 **i** 80 512 – 53 667
j 71 451 – 9764 **k** 54 632 + 17 845 + 28 783 **l** 29 677 × 47

Count to 10 000 and beyond

Copy and complete each number sequence.

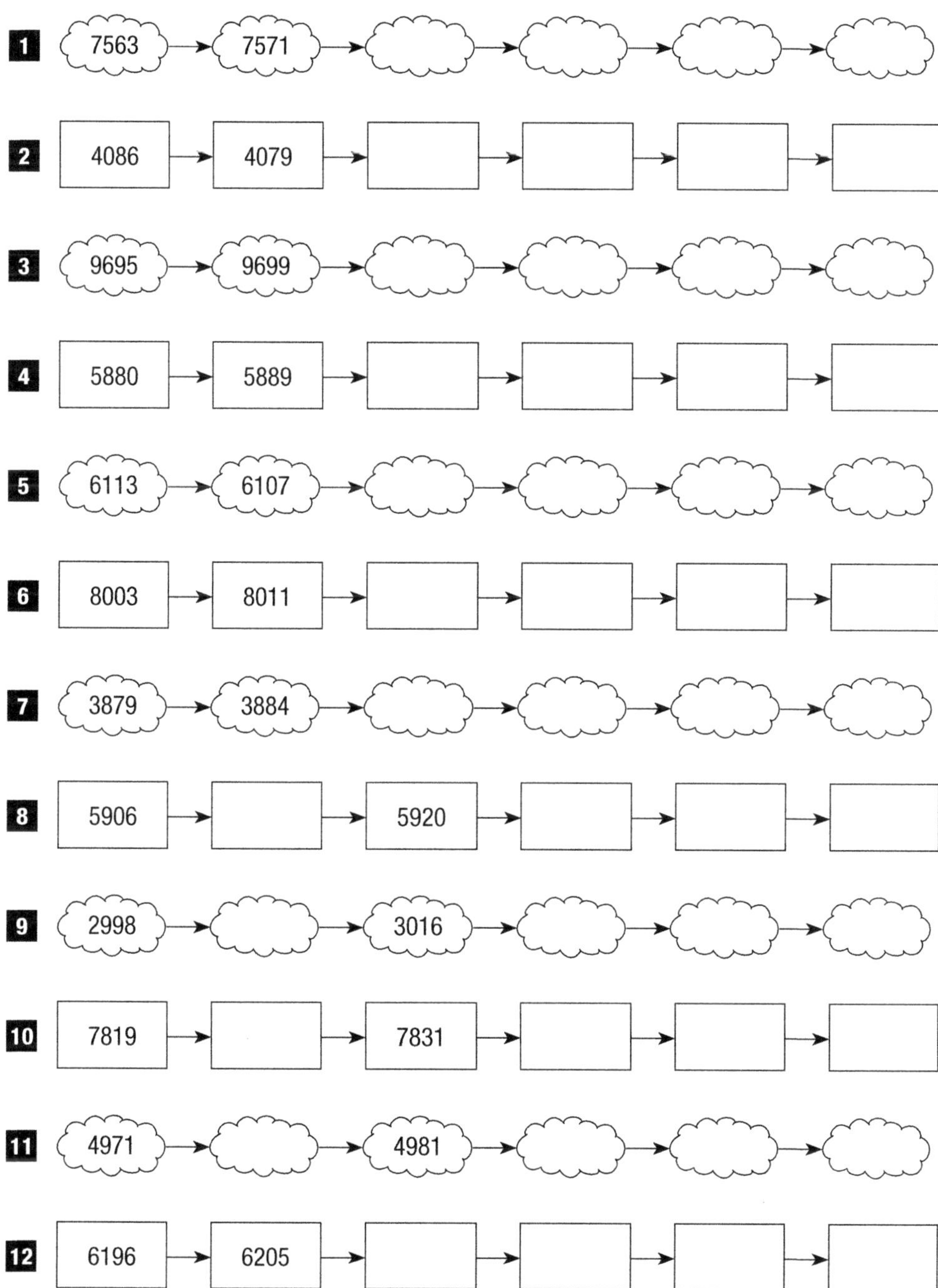

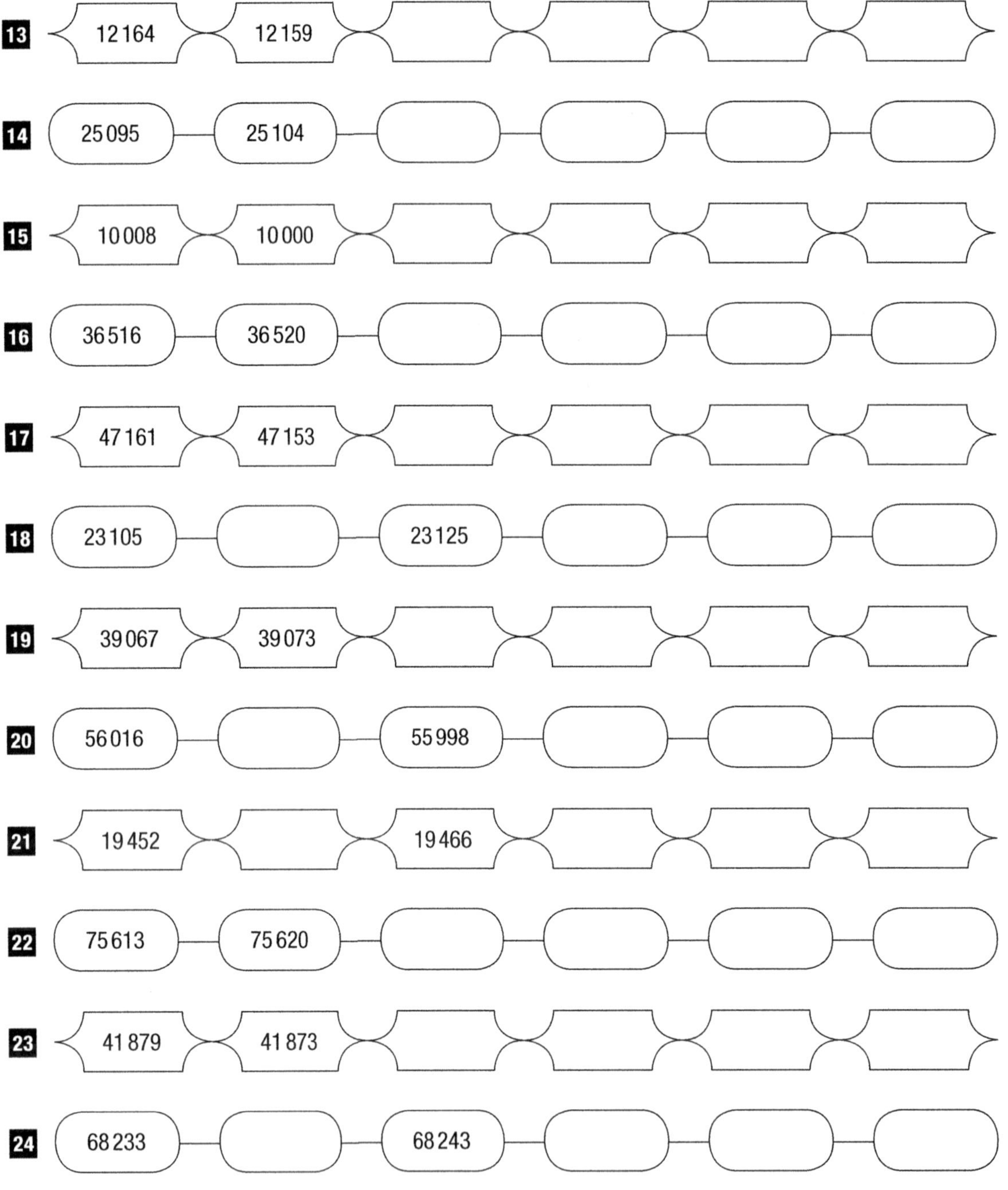

13 12 164 12 159
14 25 095 25 104
15 10 008 10 000
16 36 516 36 520
17 47 161 47 153
18 23 105 23 125
19 39 067 39 073
20 56 016 55 998
21 19 452 19 466
22 75 613 75 620
23 41 879 41 873
24 68 233 68 243

5.1.2. Apply and use the four operations to do calculations with four and five-digit numbers

Mental strategies for addition

1 Add each set of three numbers in your head and write the answer. Look for shortcuts to help you add numbers.

a 12 + 6 + 14 **b** 17 + 9 + 8 **c** 5 + 19 + 11
d 8 + 16 + 16 **e** 19 + 6 + 15 **f** 10 + 23 + 7
g 9 + 12 + 25 **h** 13 + 15 + 22 **i** 25 + 7 + 8
j 16 + 24 + 5 **k** 12 + 8 + 11 **l** 6 + 9 + 15

2 Add each set of four numbers in your head and write the answer. Look for shortcuts to help you add numbers.

a 4 + 9 + 5 + 7 **b** 10 + 6 + 12 + 8 **c** 9 + 7 + 11 + 8
d 15 + 6 + 7 + 12 **e** 18 + 4 + 20 + 6 **f** 19 + 12 + 7 + 10
g 11 + 12 + 11 + 12 **h** 8 + 11 + 13 + 9 **i** 5 + 7 + 9 + 11
j 8 + 16 + 9 + 14 **k** 20 + 9 + 30 + 7 **l** 15 + 12 + 16 + 18

3 Copy and complete each grid. Add each pair of numbers in your head.

a

+	85	78	64	103	96
9					
7					
11					
8					

b

+	126	98	75	110	87
20					
15					
19					
25					

c

+	105	93	138	112	146
14					
18					
23					
40					

d

+	99	152	106	122	163
16					
21					
33					
24					

e

+	101	97	149	58	210
19					
49					
29					
33					

f

+	197	205	161	299	119
11					
9					
51					
99					

Mental strategies for subtraction

1 Start at 500 and subtract each number progressively.

a	b	c	d	e	f
500	**500**	**500**	**500**	**500**	**500**
8	5	9	10	8	7
6	2	6	9	7	12
5	4	5	12	9	11
10	6	5	8	12	6
6	11	12	5	6	9
7	9	11	8	9	12
9	3	8	7	11	8
11	8	5	9	8	7
3	7	9	6	5	10
5	6	7	11	4	9
8	2	10	4	1	6
6	9	9	8	6	8
9	11	6	5	9	7
7	10	7	10	9	5
5	8	9	12	12	4
______	______	______	______	______	______

2 Start at the number shown and subtract each number progressively.

a	b	c	d	e	f
250	**675**	**380**	**800**	**725**	**315**
4	8	6	12	7	8
6	8	9	8	6	5
5	10	6	10	3	4
10	12	7	6	7	6
3	7	4	9	9	3
4	6	8	8	9	4
7	3	9	9	7	9
8	8	5	7	8	11
6	9	3	9	5	6
4	7	1	5	4	3
9	11	6	8	12	2
3	6	8	3	9	8
5	8	10	6	8	7
5	9	9	6	5	6
6	10	6	7	7	4
______	______	______	______	______	______

Mental strategies for multiplication and division

1 Copy and complete the multiplication grids.

a

×	6	3	7	9	4	8
5						
7						
4						
8						
2						
6						

b

×	5	2	9	11	7	12
3						
9						
7						
6						
8						
10						

2 Copy and complete the division grids.

a

÷	24	36	12	48	18	30	42
3							
6							

b

÷	32	48	16	40	72	56	24
4							
8							

3 Fill the gaps.

a

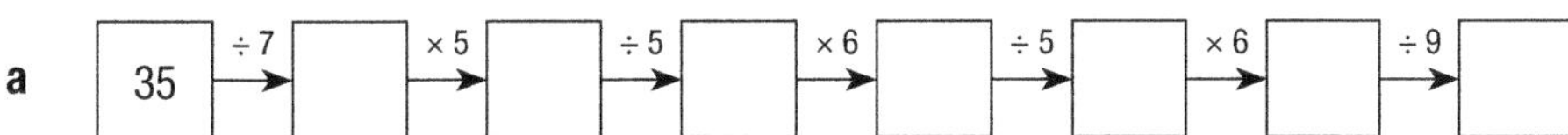

b

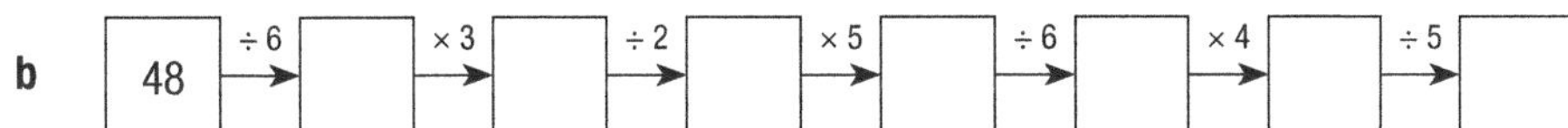

c 32 →÷ 8→ ☐ →× 4→ ☐ →÷ 8→ ☐ →× 6→ ☐ →× 4→ ☐ →÷ 8→ ☐ →× 3→ ☐

d 16 →÷ 8→ ☐ →× 12→ ☐ →÷ 6→ ☐ →× 9→ ☐ →÷ 6→ ☐ →× 8→ ☐ →÷ 12→ ☐

e 30 →÷ 5→ ☐ →× 3→ ☐ →÷ 2→ ☐ →× 6→ ☐ →÷ 9→ ☐ →× 2→ ☐ →× 3→ ☐

f 40 →÷ 4→ ☐ →× 3→ ☐ →÷ 3→ ☐ →× 2→ ☐ →÷ 5→ ☐ →× 8→ ☐ →÷ 4→ ☐

g 64 →÷ 8→ ☐ →× 2→ ☐ →÷ 4→ ☐ →× 7→ ☐ →÷ 4→ ☐ →× 7→ ☐ →÷ 7→ ☐

h 18 →÷ 6→ ☐ →× 8→ ☐ →÷ 4→ ☐ →× 2→ ☐ →× 3→ ☐ →÷ 9→ ☐ →× 8→ ☐

Use extended number facts

1 Use your knowledge of multiplication and division facts to solve these problems.

a	2 × 80	**b**	4 × 40	**c**	240 ÷ 3	**d**	140 ÷ 70
e	50 × 30	**f**	60 × 20	**g**	360 ÷ 60	**h**	400 ÷ 5
i	70 × 3	**j**	30 × 30	**k**	490 ÷ 70	**l**	560 ÷ 8
m	420 ÷ 60	**n**	40 × 50	**o**	30 × 9	**p**	180 ÷ 60

2 Solve these problems as quickly as you can. Ask a friend to time you.

a	540 ÷ 90	**b**	900 × 8	**c**	450 ÷ 9	**d**	70 × 70
e	8 × 600	**f**	280 ÷ 40	**g**	30 × 40	**h**	640 ÷ 8
i	810 ÷ 90	**j**	6 × 600	**k**	80 × 7	**l**	720 ÷ 80
m	210 ÷ 3	**n**	40 × 60	**o**	5 × 500	**p**	500 ÷ 50

3 Use your knowledge of addition and subtraction facts to solve these problems.

a	60 + 70	**b**	80 – 50	**c**	90 + 30	**d**	70 – 30
e	80 – 60	**f**	80 + 80	**g**	90 + 20	**h**	110 – 70
i	50 + 70	**j**	60 – 30	**k**	120 – 40	**l**	90 + 80
m	90 – 70	**n**	130 – 50	**o**	70 + 40	**p**	50 + 90

4 Solve these problems as quickly as you can. Ask a friend to time you.

a	80 + 30	**b**	50 + 110	**c**	90 – 60	**d**	80 + 70
e	140 – 70	**f**	120 – 30	**g**	40 + 80	**h**	20 + 90
i	110 – 30	**j**	150 – 60	**k**	80 + 20	**l**	50 + 60
m	100 – 70	**n**	120 – 80	**o**	110 – 90	**p**	110 + 70

5 Copy and complete this chart.

×	7	4	8	6	9	11	5	10	3	12
30										
50										
90										
40										
70										
80										

Use extended number facts for multiplication

Play this game with a friend.

Each player needs a set of counters (each set a different colour).

Take turns to choose two numbers, one from each cloud.

Multiply the two numbers together and if the answer is on the grid, place a counter on this number. Only one counter may be placed on each number.

The first player to get four counters in a row (in any direction – vertical, horizontal or diagonal) wins the game.

480	540	350	140
720	300	240	450
400	150	630	270
640	560	320	490

20
70
40
90
30
80
50

Use mental strategies for the four processes

1 Which totals will be more than 1000? Estimate each addition first and then work them out to see if you were correct.

a	b	c	d	e	f
156	310	267	275	192	321
370	176	326	180	365	497
225	108	185	107	289	116
+ 496	+ 229	+ 593	+ 249	+ 446	+ 182
____	____	____	____	____	____

2 Which totals will be more than 2000? Estimate each addition first and then work them out to see if you were correct.

a	b	c	d	e	f
536	726	642	891	617	269
275	885	354	723	148	306
618	316	713	567	377	357
+ 492	+ 450	+ 238	+ 536	+ 634	+ 518
____	____	____	____	____	____

3 Do these addition and subtraction calculations in your head.

a	56 + 53	**b**	220 + 70	**c**	500 – 87	**d**	139 – 50
e	850 + 400	**f**	745 – 47	**g**	600 + 500	**h**	5500 + 4000
i	839 – 600	**j**	674 – 99	**k**	1000 – 475	**l**	538 – 139
m	3600 + 800	**n**	614 – 15	**o**	487 + 300	**p**	2900 – 500
q	513 – 199	**r**	201 + 362	**s**	250 + 750	**t**	697 + 99
u	647 – 102	**v**	198 + 401	**w**	715 – 299	**x**	379 + 98

4 Do these multiplication and division calculations in your head.

a	25 × 4	**b**	750 ÷ 10	**c**	30 × 30	**d**	299 × 3
e	930 ÷ 93	**f**	320 ÷ 8	**g**	71 × 100	**h**	5 × 80
i	15 000 ÷ 3000	**j**	6 × 501	**k**	650 × 2	**l**	720 ÷ 80
m	8000 ÷ 200	**n**	36 × 20	**o**	5 × 199	**p**	7600 ÷ 76
q	500 ÷ 25	**r**	59 × 6	**s**	25 × 12	**t**	7000 ÷ 500
u	3000 ÷ 200	**v**	4500 ÷ 450	**w**	498 × 4	**x**	40 × 250

5 Write the calculations that you estimate to have an answer of more than 1000.

a	3726 ÷ 38	**b**	105 × 12	**c**	863 + 211	**d**	1373 – 250
e	25 000 ÷ 11	**f**	25 × 24	**g**	156 + 780 + 25	**h**	2560 – 1600
i	5675 ÷ 50	**j**	99 × 11	**k**	538 + 167 + 412	**l**	4107 – 2500

Use symbols and words for the four processes

Solve each set of problems. Ask a friend to time you on the last two sets.

1
- a 38 minus 17 = ______
- b 19 more than 42 = ______
- c 18 + 15 + 12 = ______
- d Sum of 36 and 16 = ______
- e 55 divided by 5 = ______
- f 8 times 25 = ______
- g Difference between 33 and 16 = ______
- h 48 + ______ = 80
- i Product of 8 and 4 = ______
- j Subtract 25 from 64 = ______
- k 18 plus 46 = ______
- l Quotient of 45 and 9 = ______
- m 95 – ______ = 74
- n 23 + 16 + 17 = ______
- o 6 times 50 = ______
- p Double 36 = ______

2
- a How many 4s in 48? ______
- b 2 times 75 = ______
- c 35 + 45 = ______
- d 31 + ______ = 80
- e Difference between 47 and 18 = ______
- f 54 minus 16 = ______
- g Product of 7 and 5 = ______
- h Subtract 13 from 52 = ______
- i ______ + 63 = 100
- j 64 divided by 8 = ______
- k 92 take away 25 = ______
- l 3 times 150 = ______
- m 44 plus 46 = ______
- n Quotient of 81 and 9 = ______
- o 28 + 14 + 22 = ______
- p Difference between 62 and 37 = ______

3
- a 75 minus 28 = ______
- b 27 + 58 = ______
- c 32 + ______ = 100
- d Product of 25 and 5 =
- e Quotient of 400 and 50 = ______
- f 6 times 19 =
- g 16 less than 74 = ______
- h 30 divided by 6 =
- i Double 48 = ______
- j Difference between 41 and 12 =
- k 12 times 25 = ______
- l 9 lots of 20 =
- m 58 + 18 = ______
- n 75 – 16 =
- o Quotient of 160 and 20 = ______
- p Share 48 among 6 =

4
- a 3 times 35 = ______
- b Half of 90 = ______
- c 96 + ______ = 130
- d Share 63 among 7 = ______
- e 5 cubed = ______
- f 34 less than 72 = ______
- g Product of 14 and 20 = ______
- h Difference between 150 and 70 = ______
- i Total of 57, 32 and 29 = ______
- j 120 minus 32 = ______
- k 4 times 26 = ______
- l 17 + 38 + 11 = ______
- m 320 divided by 8 = ______
- n Sum of 46 and 26 = ______
- o Subtract 12 from 91 = ______
- p Quotient of 100 and 4 = ______

Add four and five-digit numbers

Help Box

Add whole numbers without trading:

2564 + 5315

```
  2 5 6 4
+ 5 3 1 5
  7 8 7 9
```

Add whole numbers with trading in one or more places:

4577 + 1682

$$\begin{array}{r} 4\,5\,7\,7 \\ +{}_{1}1{}_{1}6\,8\,2 \\ \hline 6\,2\,5\,9 \end{array}$$

1 Set these additions out vertically and then solve each one.

a	3516 + 2352	**b**	4074 + 3815	**c**	2551 + 1408	**d**	5612 + 3376
e	4167 + 2375	**f**	2188 + 5657	**g**	3645 + 4787	**h**	1659 + 3482
i	15 342 + 34 587	**j**	23 518 + 17 662	**k**	35 509 + 27 286	**l**	45 118 + 29 875

Remember

When adding numbers with an unequal number of digits, keep the ones under the ones, the tens under the tens and so on.

2 Set these additions out vertically and then solve each one.

a	3779 + 289 + 4679	**b**	785 + 6759 + 308
c	5059 + 2658 + 829	**d**	2752 + 11 634 + 596
e	23 615 + 8032 + 15 086	**f**	9527 + 3605 + 12 663
g	14 509 + 6775 + 21 884	**h**	7708 + 52 617 + 29 485

3 Add the numbers horizontally and vertically to find the *super* answer for each grid.

a

1082	+	2637	+	1755	=	☐
+		+		+		+
954	+	1083	+	2751	=	☐
+		+		+		+
3665		1573		4689		☐
☐	+	☐	+	☐	=	☐

b

3644	+	2036	+	2713	=	☐
+		+		+		+
5023	+	1164	+	3082	=	☐
+		+		+		+
1620	+	2556	+	4916	=	☐
☐	+	☐	+	☐	=	☐

4 Solve these additions.

```
a   4 3 5 7 7    b   2 9 0 8 9    c   5 4 2 8 3    d   6 8 5 5 2    e   3 5 9 0 6    f   4 1 8 0 9
  + 3 8 6 1 6      + 7 7 9 5 8      + 1 9 9 3 5      + 5 4 2 0 8      + 4 8 7 5 8      + 3 7 6 3 4
  ___________      ___________      ___________      ___________      ___________      ___________
```

Subtract four and five-digit numbers

Help Box

Subtract whole numbers without trading:

4788 – 2341

$$\begin{array}{r} 4\,7\,8\,8 \\ -\,2\,3\,4\,1 \\ \hline 2\,4\,4\,7 \\ \hline \end{array}$$

Subtract whole numbers with trading in one or more places:

5273 – 2657

$$\begin{array}{r} ^{4}\not{5}\,^{1}2\,^{6}\not{7}\,^{1}3 \\ -\,2\,6\,5\,7 \\ \hline 2\,6\,1\,6 \\ \hline \end{array}$$

1 Set these subtractions out vertically and then solve each one.

a	8589 – 4136	**b**	5667 – 3426	**c**	4559 – 1427	**d**	6874 – 3622
e	5253 – 3472	**f**	9633 – 7558	**g**	6271 – 4653	**h**	4725 – 1986
i	25 627 – 18 485	**j**	27 833 – 15 748	**k**	41 246 – 28 372	**l**	37 052 – 18 717

2 Set these subtractions out vertically and then solve each one.

a	5122 – 876	**b**	7074 – 785	**c**	8163 – 647
d	6245 – 829	**e**	32 472 – 7566	**f**	25 404 – 6857
g	54 002 – 7691	**h**	47 351 – 8673	**i**	73 605 – 5128
j	41 952 – 3787	**k**	22 816 – 5495	**l**	34 780 – 7592

Remember

When subtracting numbers with unequal numbers of digits, keep the ones under the ones, the tens under the tens and so on.

3

6742	4836	11 941	15 086	12 775	8032	9817	18 938

- **a** Which two numbers have the greatest difference? What is it?
- **b** Which two numbers have a difference of 6163?
- **c** Which two numbers have a difference of 7054?
- **d** Which two numbers have a difference of less than 1000?
- **e** Find three pairs of numbers that have a difference between 2000 and 3000.

4 Help the bank work out how much money is left in each account.

a

Withdrawal	Balance
	K7638
K419	
K1567	
K832	
K1236	
K1139	
K594	

b

Withdrawal	Balance
	K14 500
K1063	
K1644	
K969	
K930	
K2065	
K1337	

Multiply four and five-digit numbers

Help Box

Short multiplication

Multiply whole numbers by a single digit:

$$\begin{array}{r} {}^{2}3\,{}^{1}5\,{}^{3}2\,7 \\ \times\ 5 \\ \hline 1\,7\,6\,3\,5 \\ \hline \end{array}$$

Long multiplication

Multiply whole numbers by two-digit numbers:

$$\begin{array}{r} {}^{3}_{1}2\,{}^{2}7\,{}^{2}_{1}4\,5 \\ \times\ 2\,5 \\ \hline 1\,3\,7\,2\,5 \\ +\ 5\,4\,9\,0\,0 \\ \hline 6\,8\,6\,2\,5 \\ \hline \end{array}$$

1 Work out the answers to find the lucky number.

LUCKY NUMBER 15 428

a 1654 × 4

b 2503 × 6

c 1826 × 3

d 4673 × 2

e 2756 × 4

f 1967 × 5

g 3524 × 3

h 5014 × 4

i 1367 × 5

j 3857 × 4

k 4025 × 3

l 3660 × 6

m 2149 × 3

n 2933 × 7

2 Set out and solve these long multiplications.

a 752 × 18 b 673 × 24 c 485 × 19 d 691 × 25

e 4352 × 23 f 7165 × 34 g 13 442 × 17 h 27 815 × 28

3 Work out these multiplications and then write the letters that are above the five largest products. If you write them in order starting with the letter above the largest product, you will find the name of a town in Papua New Guinea.

a **M** 2137 × 32

b **D** 1864 × 26

c **R** 3478 × 19

d **P** 2873 × 33

e **T** 5126 × 29

f **I** 4972 × 45

g **S** 3067 × 17

h **B** 2336 × 37

i **G** 4072 × 24

j **E** 3891 × 43

k **K** 6012 × 39

l **A** 5493 × 22

Divide four and five-digit numbers

Help Box

Short division

Divide whole numbers by a single digit:

4278 ÷ 5 =

```
   8 5 5 remainder 3
5)4 2²7²8
```

Answer = 855 remainder 3

or $855\frac{3}{5}$

Long division

Divide whole numbers by two-digit numbers:

2789 ÷ 15 =

```
       1 8 5
1 5)2 7 8 9
  - 1 5
    1 2 8
  - 1 2 0
        8 9
      - 7 5
        1 4
```

Answer = 185 remainder 14

or $185\frac{14}{15}$

1 Set out and solve these divisions. Write any remainders as fractions.

a	2683 ÷ 5	b	7512 ÷ 4	c	5706 ÷ 7	d	3065 ÷ 2
e	4673 ÷ 8	f	3689 ÷ 6	g	5109 ÷ 3	h	2927 ÷ 4
i	40 315 ÷ 6	j	59 692 ÷ 8	k	25 516 ÷ 3	l	47 061 ÷ 9

2 Solve these divisions to see how many fish you can catch.

a 

b 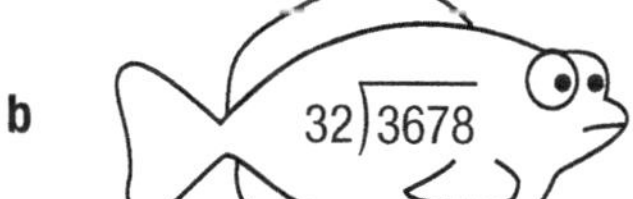

c 

d 

e 

f 

g 

h 

i 

How many fish did you catch?

5.1.3 Use the four operations to solve problems related to proper fractions

Find fractional parts of groups

Help Box

Unit fractions (fractions with a numerator of 1):

$\frac{1}{5}$ of 20 = 20 ÷ 5 = 4

Fractions with a numerator other than 1, for example $\frac{2}{3}$ of 15:

1. Find $\frac{1}{3}$ of 15: 15 ÷ 3 = 5
2. Multiply 5 by 2 to find $\frac{2}{3}$: 5 × 2 = 10, so $\frac{2}{3}$ of 15 = 10

1 Find the fraction of each group.

a	$\frac{1}{2}$ of 16	**b**	$\frac{1}{2}$ of 20	**c**	$\frac{1}{4}$ of 8	**d**	$\frac{1}{4}$ of 12
e	$\frac{1}{4}$ of 20	**f**	$\frac{1}{3}$ of 6	**g**	$\frac{1}{3}$ of 15	**h**	$\frac{1}{5}$ of 15
i	$\frac{1}{5}$ of 5	**j**	$\frac{1}{5}$ of 10	**k**	$\frac{1}{2}$ of 4	**l**	$\frac{1}{2}$ of 14
m	$\frac{1}{3}$ of 24	**n**	$\frac{1}{4}$ of 4	**o**	$\frac{1}{10}$ of 20	**p**	$\frac{1}{10}$ of 50
q	$\frac{1}{8}$ of 16	**r**	$\frac{1}{6}$ of 12	**s**	$\frac{1}{3}$ of 9	**t**	$\frac{1}{6}$ of 30
u	$\frac{1}{6}$ of 18	**v**	$\frac{1}{3}$ of 18	**w**	$\frac{1}{4}$ of 40	**x**	$\frac{1}{4}$ of 32

2 Find the fraction of each group.

a	$\frac{2}{3}$ of 12	**b**	$\frac{3}{4}$ of 16	**c**	$\frac{2}{3}$ of 30	**d**	$\frac{2}{5}$ of 20
e	$\frac{3}{5}$ of 20	**f**	$\frac{4}{5}$ of 20	**g**	$\frac{5}{6}$ of 36	**h**	$\frac{3}{10}$ of 30
i	$\frac{7}{10}$ of 30	**j**	$\frac{3}{8}$ of 24	**k**	$\frac{7}{8}$ of 24	**l**	$\frac{2}{5}$ of 25
m	$\frac{3}{4}$ of 20	**n**	$\frac{7}{10}$ of 50	**o**	$\frac{4}{5}$ of 30	**p**	$\frac{5}{8}$ of 32
q	$\frac{2}{9}$ of 18	**r**	$\frac{2}{3}$ of 27	**s**	$\frac{2}{5}$ of 35	**t**	$\frac{5}{6}$ of 12
u	$\frac{3}{5}$ of 15	**v**	$\frac{2}{7}$ of 14	**w**	$\frac{3}{4}$ of 28	**x**	$\frac{7}{8}$ of 32

3 Solve these problems.

a There are 24 people. $\frac{1}{4}$ of the people are male. How many are female?

b There are 16 cats and dogs in total. $\frac{3}{4}$ are cats. How many are dogs?

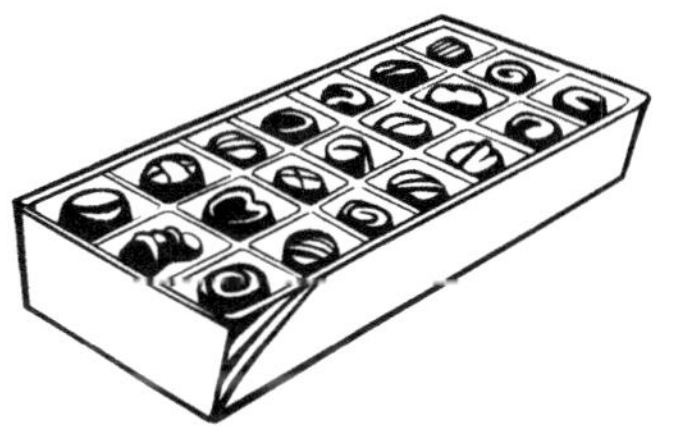

c $\frac{2}{3}$ of 24 chocolates have hard centres. How many have soft centres?

d $\frac{4}{5}$ of a group of 30 animals are goats. How many animals are not goats?

e There are 32 cakes. $\frac{5}{8}$ of them have chocolate icing. How many is this?

f Rani has 18 coins. $\frac{4}{9}$ of them are gold. How many are not gold?

g 36 people are waiting for a PMV. $\frac{5}{6}$ of these people are adults. How many are not adults?

h There are 10 cans of tinned fish. Regina opens $\frac{3}{5}$ of them. How many is this?

i There are 21 pieces of fruit. $\frac{4}{7}$ of these are bananas. How many are not bananas?

j A class has 28 students. $\frac{5}{7}$ of them like to read. How many do not like to read?

4 Try these more difficult problems.

a $\frac{3}{4}$ of 100	b $\frac{2}{3}$ of 120	c $\frac{4}{5}$ of 70	d $\frac{2}{5}$ of 200
e $\frac{7}{8}$ of 1000	f $\frac{3}{4}$ of 1000	g $\frac{5}{6}$ of 90	h $\frac{2}{7}$ of 280
i $\frac{5}{9}$ of 630	j $\frac{3}{8}$ of 160	k $\frac{3}{5}$ of 500	l $\frac{5}{8}$ of 2000
m $\frac{6}{7}$ of 105	n $\frac{4}{5}$ of 800	o $\frac{3}{7}$ of 350	p $\frac{2}{3}$ of 1800
q $\frac{5}{7}$ of 140	r $\frac{4}{9}$ of 360	s $\frac{7}{10}$ of 5000	t $\frac{7}{8}$ of 640
u $\frac{2}{9}$ of 810	v $\frac{5}{6}$ of 2400	w $\frac{7}{9}$ of 2700	x $\frac{3}{10}$ of 380

Mixed numbers, proper and improper fractions

Help Box

An improper fraction is a fraction with a numerator that is greater than or equal to the denominator. For example, $\frac{7}{3}$, $\frac{12}{7}$ and $\frac{4}{4}$ are improper fractions.

A proper fraction is a fraction with a numerator that is less than the denominator. For example, $\frac{1}{5}$, $\frac{7}{8}$ and $\frac{15}{100}$ are proper fractions.

A mixed number is a number that has a whole number and a proper fraction. For example, $3\frac{1}{2}$, $7\frac{1}{3}$ and $5\frac{2}{7}$ are mixed numbers.

$\frac{1}{4}$

$\frac{1}{4}$ $\frac{7}{7}$ $\frac{9}{7}$ $\frac{12}{5}$ $\frac{1}{3}$ $1\frac{1}{2}$ $\frac{10}{7}$

$\frac{11}{5}$ $6\frac{2}{3}$ $\frac{1}{5}$ $\frac{1}{2}$ $\frac{3}{4}$ $\frac{3}{3}$ $2\frac{3}{4}$ $\frac{10}{10}$

1 From the group of fractions above, write:

a	the proper fractions	**b**	the mixed numbers	**c**	the improper fractions
d	fractions equal to 1	**e**	fractions larger than 1	**f**	fractions smaller than 1

2 Convert these improper fractions to mixed numbers.

Remember

To convert an improper fraction to a mixed number, divide the numerator by the denominator.

$\frac{16}{9} = 16 \div 9 = 1\frac{7}{9}$

a	$\frac{9}{2}$	**b**	$\frac{17}{5}$	**c**	$\frac{14}{3}$	**d**	$\frac{21}{4}$
e	$\frac{8}{3}$	**f**	$\frac{10}{3}$	**g**	$\frac{19}{6}$	**h**	$\frac{23}{7}$
i	$\frac{15}{8}$	**j**	$\frac{20}{7}$	**k**	$\frac{13}{4}$	**l**	$\frac{22}{9}$
m	$\frac{16}{3}$	**n**	$\frac{11}{7}$	**o**	$\frac{17}{4}$	**p**	$\frac{25}{6}$
q	$\frac{33}{8}$	**r**	$\frac{29}{7}$	**s**	$\frac{31}{9}$	**t**	$\frac{5}{4}$
u	$\frac{21}{8}$	**v**	$\frac{35}{11}$	**w**	$\frac{9}{7}$	**x**	$\frac{24}{5}$

Remember

To convert a mixed number to an improper fraction, multiply the denominator by the whole number. Add the answer to the numerator of the fraction and place the total over the denominator.

For $2\frac{3}{7}$, multiply 2 by 7 (14), add 14 and 3 (17). Place 17 over 7: $\frac{17}{7}$

3 Convert these mixed numbers to improper fractions.

a	$2\frac{1}{4}$	**b**	$4\frac{1}{2}$	**c**	$1\frac{3}{4}$	**d**	$2\frac{5}{6}$	**e**	$1\frac{7}{8}$	**f**	$3\frac{3}{10}$
g	$5\frac{2}{3}$	**h**	$2\frac{3}{5}$	**i**	$4\frac{3}{8}$	**j**	$1\frac{4}{9}$	**k**	$2\frac{9}{11}$	**l**	$1\frac{2}{5}$
m	$3\frac{1}{6}$	**n**	$1\frac{5}{8}$	**o**	$6\frac{1}{2}$	**p**	$4\frac{1}{3}$	**q**	$3\frac{7}{9}$	**r**	$2\frac{6}{7}$
s	$4\frac{1}{10}$	**t**	$5\frac{5}{7}$	**u**	$2\frac{3}{11}$	**v**	$1\frac{8}{9}$	**w**	$6\frac{1}{7}$	**x**	$3\frac{4}{5}$

4 From the group of fractions above, write:

- **a** the fractions and mixed numbers that are between 1 and 2
- **b** the fractions and mixed numbers that are between 2 and 3
- **c** the fractions below 1
- **d** the fractions and mixed numbers above 3
- **e** the largest mixed number

Make equivalent fractions

Help Box

Equivalent fractions can be made by multiplying the numerator and the denominator by the same number.

$$\frac{1 \times 3}{5 \times 3} = \frac{3}{15}$$

so $\frac{1}{5}$ is the same as $\frac{3}{15}$.

Equivalent fractions can also be made by dividing the numerator and the denominator by the same number.

$$\frac{4 \div 2}{6 \div 2} = \frac{2}{3}$$

so $\frac{4}{6}$ is the same as $\frac{2}{3}$.

1 Work out what each fraction has been multiplied by and then write an equivalent fraction.

a $\frac{1}{6} = \frac{\square}{12}$ **b** $\frac{1}{3} = \frac{\square}{9}$ **c** $\frac{1}{7} = \frac{\square}{21}$ **d** $\frac{1}{5} = \frac{\square}{20}$

e $\frac{2}{3} = \frac{\square}{6}$ **f** $\frac{3}{5} = \frac{\square}{10}$ **g** $\frac{3}{8} = \frac{\square}{16}$ **h** $\frac{3}{4} = \frac{\square}{16}$

i $\frac{3}{7} = \frac{\square}{14}$ **j** $\frac{5}{6} = \frac{\square}{18}$ **k** $\frac{2}{9} = \frac{\square}{27}$ **l** $\frac{7}{8} = \frac{\square}{40}$

m $\frac{2}{5} = \frac{4}{\square}$ **n** $\frac{4}{7} = \frac{20}{\square}$ **o** $\frac{5}{8} = \frac{15}{\square}$ **p** $\frac{2}{3} = \frac{12}{\square}$

q $\frac{7}{9} = \frac{14}{\square}$ **r** $\frac{3}{10} = \frac{12}{\square}$ **s** $\frac{2}{11} = \frac{6}{\square}$ **t** $\frac{4}{9} = \frac{16}{\square}$

u $\frac{2}{7} = \frac{14}{\square}$ **v** $\frac{4}{5} = \frac{40}{\square}$ **w** $\frac{9}{10} = \frac{90}{\square}$ **x** $\frac{8}{11} = \frac{32}{\square}$

2 Work out what each fraction has been divided by and then write an equivalent fraction.

a $\frac{3}{9} = \frac{1}{\square}$ **b** $\frac{5}{10} = \frac{\square}{2}$ **c** $\frac{2}{8} = \frac{1}{\square}$ **d** $\frac{10}{15} = \frac{2}{\square}$

e $\frac{6}{9} = \frac{\square}{3}$ **f** $\frac{8}{12} = \frac{2}{\square}$ **g** $\frac{6}{20} = \frac{\square}{10}$ **h** $\frac{12}{15} = \frac{\square}{5}$

i $\frac{8}{10} = \frac{4}{\square}$ **j** $\frac{10}{24} = \frac{\square}{12}$ **k** $\frac{4}{16} = \frac{1}{\square}$ **l** $\frac{9}{12} = \frac{3}{\square}$

m $\frac{14}{20} = \frac{\square}{10}$ **n** $\frac{6}{10} = \frac{3}{\square}$ **o** $\frac{18}{24} = \frac{\square}{4}$ **p** $\frac{4}{12} = \frac{\square}{3}$

q $\frac{12}{16} = \frac{3}{\square}$ **r** $\frac{9}{21} = \frac{\square}{7}$ **s** $\frac{12}{30} = \frac{2}{\square}$ **t** $\frac{8}{20} = \frac{2}{\square}$

u $\frac{15}{18} = \frac{\square}{6}$ **v** $\frac{4}{22} = \frac{2}{\square}$ **w** $\frac{6}{14} = \frac{\square}{7}$ **x** $\frac{10}{25} = \frac{\square}{5}$

3 Write the fraction that is equivalent to the fraction on the left.

a $\frac{1}{4}$ $\frac{1}{2}$ $\frac{2}{8}$ $\frac{3}{10}$

b $\frac{1}{2}$ $\frac{5}{10}$ $\frac{4}{5}$ $\frac{1}{3}$

c $\frac{1}{5}$ $\frac{1}{4}$ $\frac{2}{10}$ $\frac{3}{5}$

d $\frac{4}{8}$ $\frac{1}{2}$ $\frac{6}{7}$ $\frac{2}{3}$

e $\frac{1}{3}$ $\frac{1}{6}$ $\frac{1}{4}$ $\frac{3}{9}$

f $\frac{2}{3}$ $\frac{2}{6}$ $\frac{4}{6}$ $\frac{2}{4}$

g $\frac{1}{10}$ $\frac{2}{5}$ $\frac{4}{10}$ $\frac{10}{100}$

h $\frac{3}{4}$ $\frac{6}{8}$ $\frac{2}{3}$ $\frac{1}{2}$

i $\frac{2}{4}$ $\frac{1}{2}$ $\frac{6}{8}$ $\frac{2}{3}$

j $\frac{1}{6}$ $\frac{2}{5}$ $\frac{2}{12}$ $\frac{1}{4}$

4 Write the fractions that are equivalent to the fraction on the left.

a $\frac{1}{2}$ $\frac{5}{10}$ $\frac{6}{12}$ $\frac{4}{5}$

b $\frac{2}{3}$ $\frac{1}{5}$ $\frac{4}{6}$ $\frac{6}{10}$

c $\frac{3}{6}$ $\frac{6}{12}$ $\frac{4}{6}$ $\frac{8}{16}$

d $\frac{1}{3}$ $\frac{3}{4}$ $\frac{3}{9}$ $\frac{2}{6}$

e $\frac{3}{4}$ $\frac{4}{8}$ $\frac{2}{3}$ $\frac{6}{8}$

f $\frac{8}{10}$ $\frac{4}{5}$ $\frac{3}{4}$ $\frac{1}{5}$

g $\frac{4}{16}$ $\frac{1}{4}$ $\frac{2}{8}$ $\frac{3}{12}$

h $\frac{2}{10}$ $\frac{1}{2}$ $\frac{5}{10}$ $\frac{1}{5}$

i $\frac{1}{4}$ $\frac{10}{40}$ $\frac{4}{1}$ $\frac{3}{4}$

j $\frac{3}{3}$ $\frac{5}{6}$ $\frac{4}{4}$ $\frac{9}{9}$

5 = or ≠ ?

a $\frac{2}{3}$ ☐ $\frac{2}{5}$ b $\frac{1}{8}$ ☐ $\frac{2}{4}$ c $\frac{3}{4}$ ☐ $\frac{15}{20}$ d $\frac{5}{6}$ ☐ $\frac{30}{36}$

e $\frac{24}{28}$ ☐ $\frac{6}{7}$ f $\frac{4}{7}$ ☐ $\frac{21}{12}$ g $\frac{8}{12}$ ☐ $\frac{2}{3}$ h $\frac{2}{7}$ ☐ $\frac{4}{11}$

i $\frac{30}{50}$ ☐ $\frac{4}{5}$ j $\frac{7}{9}$ ☐ $\frac{35}{45}$ k $\frac{9}{10}$ ☐ $\frac{4}{5}$ l $\frac{5}{8}$ ☐ $\frac{50}{80}$

m $\frac{25}{30}$ ☐ $\frac{5}{6}$ n $\frac{4}{9}$ ☐ $\frac{8}{21}$ o $\frac{16}{18}$ ☐ $\frac{8}{9}$ p $\frac{3}{4}$ ☐ $\frac{2}{3}$

q $\frac{700}{1000}$ ☐ $\frac{7}{10}$ r $\frac{75}{100}$ ☐ $\frac{3}{4}$ s $\frac{35}{70}$ ☐ $\frac{1}{2}$ t $\frac{8}{11}$ ☐ $\frac{16}{25}$

u $\frac{1}{4}$ ☐ $\frac{12}{48}$ v $\frac{3}{12}$ ☐ $\frac{1}{3}$ w $\frac{5}{9}$ ☐ $\frac{15}{27}$ x $\frac{250}{1000}$ ☐ $\frac{1}{4}$

Add and subtract fractions

Help Box

Fractions with like denominators:

- $\frac{3}{10} + \frac{6}{10} = \frac{9}{10}$
- $\frac{11}{12} - \frac{4}{12} = \frac{7}{12}$

Fractions with similar denominators:

- $\frac{3}{8} + \frac{1}{4} = \frac{3}{8} + \frac{2}{8} = \frac{5}{8}$
- $\frac{8}{9} - \frac{2}{3} = \frac{8}{9} - \frac{6}{9} = \frac{2}{9}$

1 These additions and subtractions have like denominators. Convert any answers that are improper fractions to mixed numbers. Write all answers in their simplest form.

a $\frac{1}{6} + \frac{4}{6}$ **b** $\frac{3}{8} + \frac{2}{8}$ **c** $\frac{3}{7} + \frac{3}{7}$ **d** $\frac{9}{10} - \frac{6}{10}$

e $\frac{7}{8} - \frac{1}{8}$ **f** $\frac{11}{12} - \frac{4}{12}$ **g** $\frac{3}{10} + \frac{6}{10}$ **h** $\frac{9}{12} - \frac{4}{12}$

i $\frac{4}{5} - \frac{2}{5}$ **j** $\frac{4}{8} + \frac{3}{8}$ **k** $\frac{5}{6} + \frac{5}{6}$ **l** $\frac{8}{10} + \frac{5}{10}$

m $\frac{2}{5} + \frac{4}{5}$ **n** $\frac{9}{11} - \frac{3}{11}$ **o** $\frac{9}{15} - \frac{7}{15}$ **p** $\frac{3}{4} + \frac{3}{4}$

q $\frac{6}{7} - \frac{2}{7}$ **r** $\frac{4}{9} + \frac{7}{9}$ **s** $\frac{17}{25} - \frac{8}{25}$ **t** $\frac{7}{10} + \frac{8}{10}$

u $\frac{13}{20} + \frac{15}{20}$ **v** $\frac{13}{16} - \frac{7}{16}$ **w** $\frac{7}{18} + \frac{15}{18}$ **x** $\frac{14}{15} - \frac{6}{15}$

2 These additions and subtractions have similar denominators. First make sure the fractions have the same denominator and then add or subtract them. Write all answers in their simplest form.

a $\frac{5}{12} + \frac{5}{6}$ **b** $\frac{2}{3} + \frac{4}{9}$ **c** $\frac{11}{12} - \frac{1}{6}$ **d** $\frac{9}{10} - \frac{2}{5}$

e $\frac{17}{20} - \frac{3}{10}$ **f** $\frac{11}{15} - \frac{3}{5}$ **g** $\frac{7}{8} + \frac{3}{4}$ **h** $\frac{1}{2} + \frac{5}{8}$

i $\frac{13}{16} + \frac{5}{8}$ **j** $\frac{5}{12} + \frac{2}{3}$ **k** $\frac{11}{14} - \frac{2}{7}$ **l** $\frac{7}{12} - \frac{1}{3}$

m $\frac{14}{15} - \frac{2}{5}$ **n** $\frac{21}{25} - \frac{4}{5}$ **o** $\frac{3}{8} + \frac{5}{24}$ **p** $\frac{11}{18} + \frac{5}{6}$

q $\frac{3}{4} + \frac{7}{12}$ **r** $\frac{7}{10} + \frac{1}{2}$ **s** $\frac{15}{16} - \frac{3}{4}$ **t** $\frac{13}{21} - \frac{3}{7}$

u $\frac{19}{22} - \frac{5}{11}$ **v** $\frac{23}{30} - \frac{3}{10}$ **w** $\frac{7}{25} + \frac{4}{5}$ **x** $\frac{2}{3} + \frac{7}{24}$

3 Write the correct answer for each addition or subtraction.

a $\frac{4}{5}+\frac{2}{10}=$ $\frac{6}{10}$ 1 $\frac{6}{15}$

b $\frac{7}{8}-\frac{1}{4}=$ $\frac{5}{8}$ $\frac{6}{4}$ $\frac{6}{8}$

c $\frac{9}{10}-\frac{1}{5}=$ $\frac{8}{5}$ $\frac{4}{5}$ $\frac{7}{10}$

d $\frac{5}{6}+\frac{2}{3}=$ $\frac{7}{9}$ $1\frac{3}{6}$ $\frac{4}{3}$

e $\frac{1}{2}+\frac{5}{8}=$ $1\frac{1}{8}$ $\frac{6}{10}$ 1

f $\frac{3}{4}-\frac{1}{8}=$ $\frac{2}{6}$ $\frac{5}{8}$ $\frac{1}{2}$

g $\frac{7}{10}-\frac{1}{2}=$ $\frac{2}{10}$ $\frac{6}{8}$ $\frac{2}{5}$

h $\frac{1}{2}+\frac{3}{6}=$ $\frac{4}{8}$ $\frac{3}{4}$ 1

i $\frac{5}{12}+\frac{1}{6}=$ $\frac{6}{18}$ $\frac{7}{12}$ $\frac{6}{12}$

j $\frac{8}{9}-\frac{2}{3}=$ $\frac{6}{9}$ $\frac{2}{9}$ $\frac{6}{6}$

k $\frac{5}{6}-\frac{7}{12}=$ $\frac{3}{4}$ $\frac{1}{4}$ $\frac{5}{12}$

l $\frac{11}{12}+\frac{5}{6}=$ $1\frac{3}{4}$ $1\frac{1}{2}$ $1\frac{1}{4}$

m $\frac{7}{16}+\frac{3}{4}=$ $1\frac{3}{16}$ $1\frac{1}{4}$ $\frac{11}{16}$

n $\frac{13}{15}-\frac{3}{5}=$ $\frac{10}{15}$ $\frac{3}{10}$ $\frac{4}{15}$

4 Copy and complete the grids.

a

+	$\frac{1}{4}$	$\frac{1}{2}$	$\frac{1}{8}$	$\frac{3}{4}$	$\frac{3}{8}$	1
$\frac{1}{4}$						
$\frac{1}{2}$						

b

−	$\frac{1}{2}$	1	$\frac{3}{4}$	$\frac{7}{8}$	2	$\frac{5}{8}$
$\frac{1}{2}$						
$\frac{1}{4}$						

5 Copy and fill the spaces. Write all answers in their simplest form.

a $\frac{1}{3}+\square=1$

b $\frac{5}{6}+\square=1$

c $\frac{3}{8}+\square=1$

d $\frac{9}{10}+\square=2$

e $1-\square=\frac{1}{4}$

f $\frac{3}{4}-\square=\frac{5}{8}$

g $\frac{7}{8}-\square=\frac{1}{4}$

h $\frac{11}{12}-\square=\frac{1}{2}$

i $\frac{4}{15}+\square=\frac{2}{3}$

j $\frac{17}{20}-\square=\frac{3}{5}$

k $\frac{3}{10}+\square=\frac{4}{5}$

l $\frac{8}{9}-\square=\frac{1}{3}$

m $1\frac{2}{3}+\square=5$

n $2\frac{4}{7}+\square=6$

o $5-\square=2\frac{3}{5}$

p $8-\square=3\frac{1}{6}$

q $\frac{7}{18}-\square=\frac{2}{9}$

r $\frac{11}{12}-\square=\frac{3}{4}$

s $\frac{4}{25}+\square=\frac{14}{25}$

t $\frac{5}{8}+\square=1\frac{3}{8}$

u $\frac{2}{3}+\square=1\frac{5}{9}$

v $\frac{11}{16}-\square=\frac{5}{16}$

w $\frac{11}{12}+\square=1\frac{1}{4}$

x $5\frac{1}{2}-\square=2\frac{1}{8}$

Add and subtract mixed numbers

Help Box

Addition and subtraction:
Make sure the fractions have the same denominators.
The whole numbers can be added or subtracted separately.

$$2\frac{2}{3} + 3\frac{5}{6} = 5 + (\frac{2}{3} + \frac{5}{6})$$
$$= 5 + (\frac{4}{6} + \frac{5}{6})$$
$$= 5 + \frac{9}{6}$$
$$= 5 + 1\frac{3}{6}$$
$$= 6\frac{3}{6} \text{ or } 6\frac{1}{2}$$

$$4\frac{8}{9} - 2\frac{1}{3} = 4\frac{8}{9} - 2\frac{3}{9}$$
$$= (4 - 2) + (\frac{8}{9} - \frac{3}{9})$$
$$= 2\frac{5}{9}$$

Add or subtract these mixed numbers to see how many goals you can kick.

1

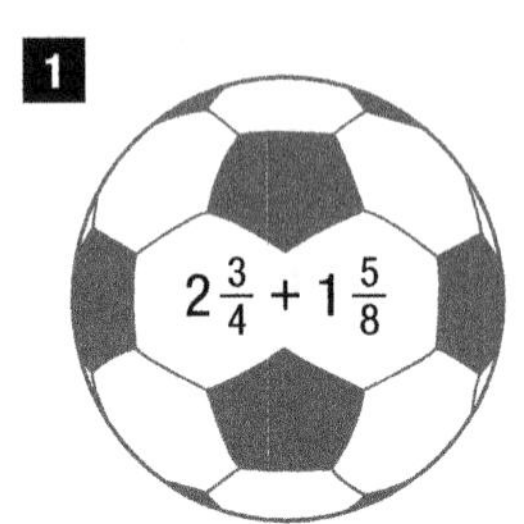

2

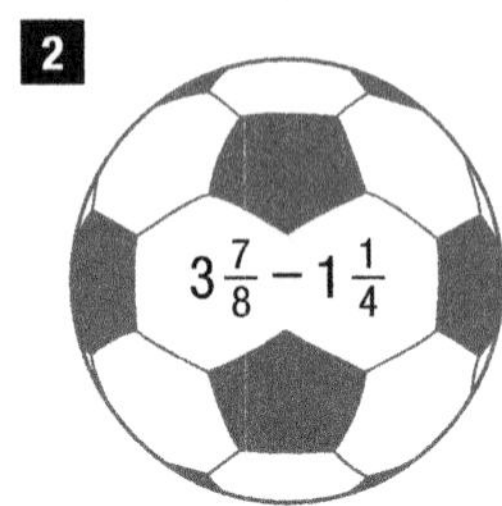

3

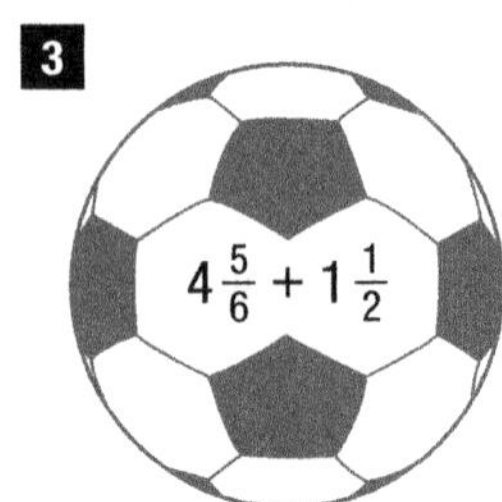

4

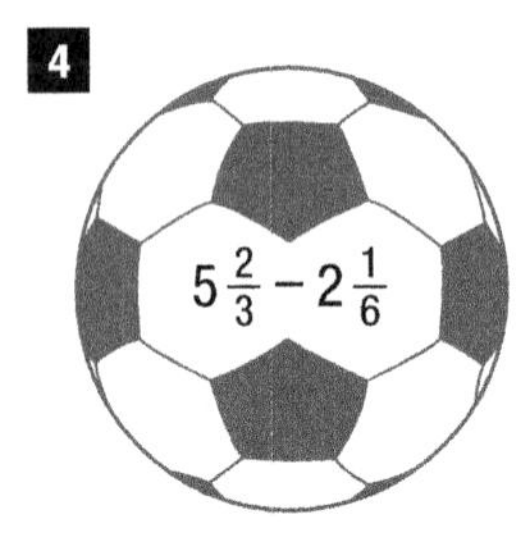

5

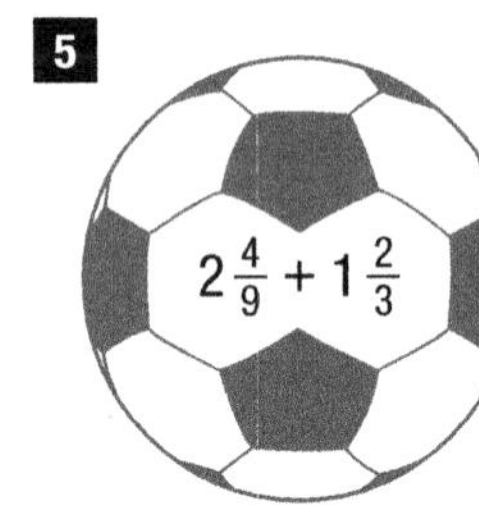

6

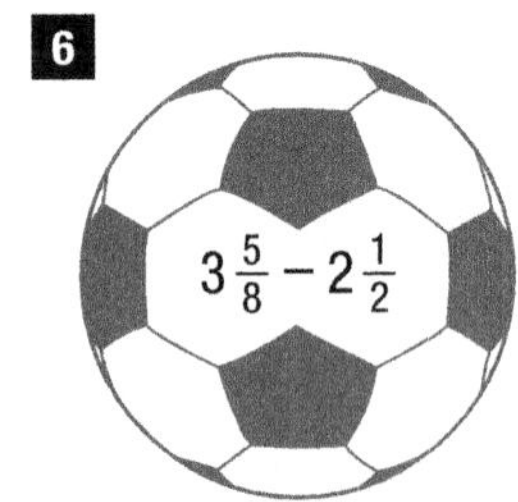

7

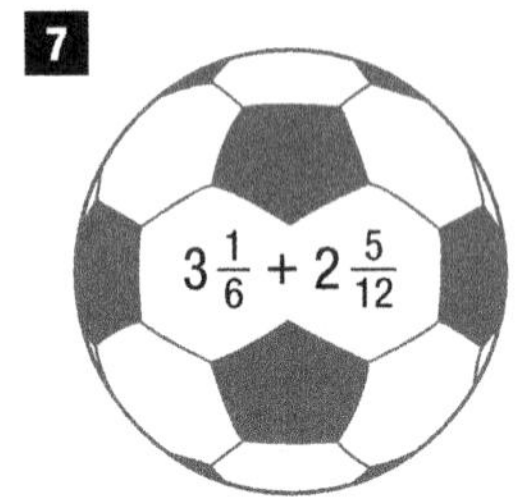

8

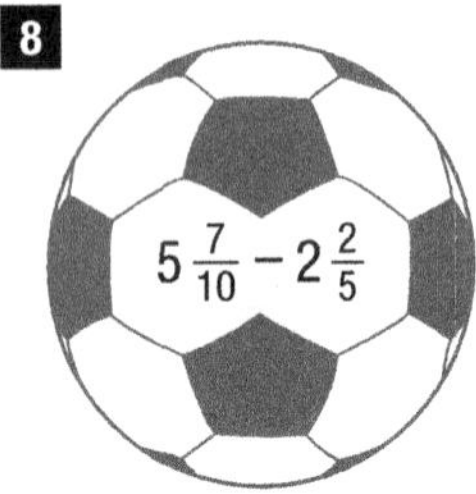

9

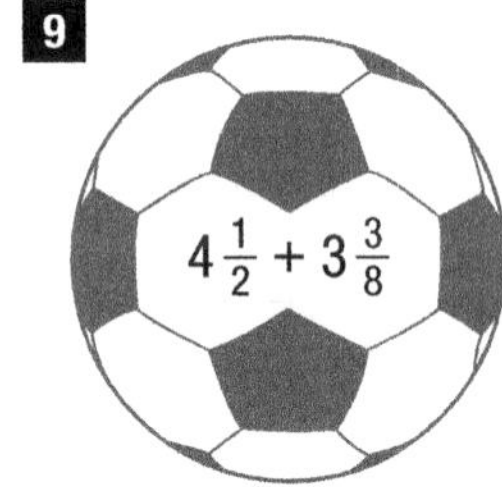

10

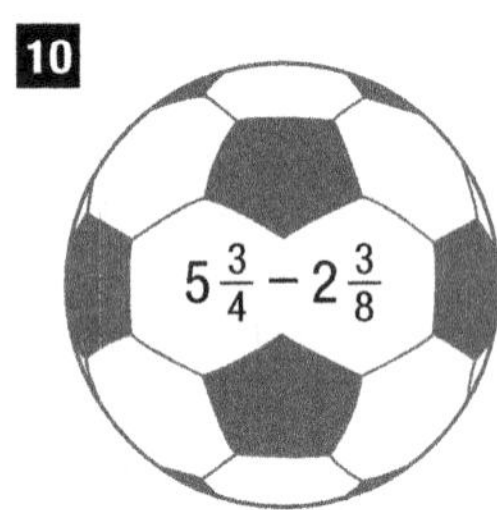

11

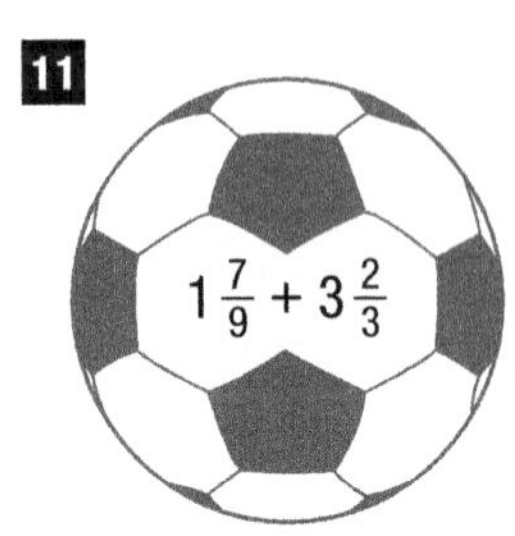

12

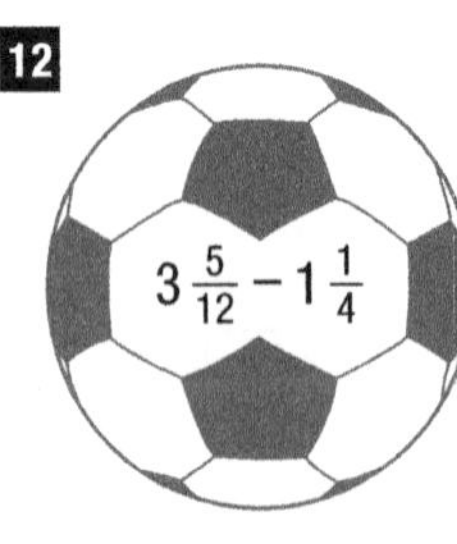

13

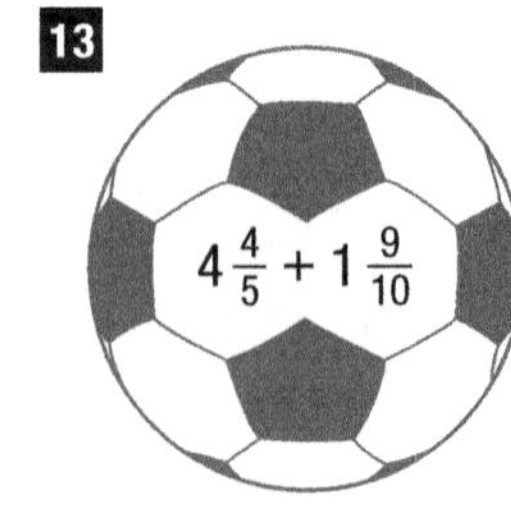

14

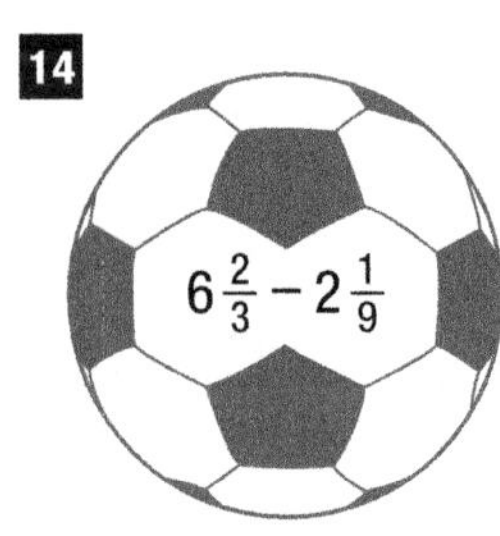

How many goals did you kick?

5.1.4 Use the four operations to solve problems related to decimals

Place value with decimals

1 What numbers are these?

a	6 tenths, 8 ones, 3 hundredths	**b**	7 ones, 5 hundredths, 1 tenth
c	4 tenths, 3 ones, 8 tens	**d**	9 tens, 6 hundredths, 2 ones
e	2 ones, 7 tenths, 1 ten	**f**	4 tenths, 8 tens
g	5 tens, 4 tenths, 6 ones	**h**	3 ones, 9 hundredths, 1 tenth
i	2 hundreds, 6 hundredths, 3 ones	**j**	7 hundredths, 4 tenths, 8 ones

2 Write the value of the underlined digit.

a	$3\underline{6}.5$	**b**	$7.\underline{9}2$	**c**	$1\underline{5}3.08$	**d**	$58.\underline{6}$	**e**	$172.7\underline{5}$	**f**	$34.\underline{4}7$
g	$2.6\underline{7}5$	**h**	$16.\underline{0}9$	**i**	$\underline{2}95.3$	**j**	$41.\underline{7}36$	**k**	$7.\underline{8}$	**l**	$30.9\underline{6}$
m	$8.65\underline{2}$	**n**	$3\underline{9}.712$	**o**	$65.0\underline{8}1$	**p**	$12.07\underline{6}$	**q**	$17.\underline{9}3$	**r**	$\underline{5}.067$

3 Write the number from each group in which 5 is worth the most.

a	3.52	56.7	1.25	**b**	6.715	35.71	2.563
c	4.57	62.35	43.265	**d**	18.05	56.34	15.12

4 Write the number from each group in which 8 is worth the least.

a	28.66	31.82	596.18	**b**	81.3	79.8	68.7
c	5.208	98.5	37.08	**d**	24.58	367.84	4.018

5 Copy and complete these.

a $25.8 = (2 \times \square) + (5 \times \square) + (8 \times \frac{\square}{\square})$

b $3.67 = (3 \times \square) + (6 \times \frac{\square}{\square}) + (7 \times \frac{\square}{\square})$

c $12.3 = (1 \times \square) + (2 \times \square) + (3 \times \frac{\square}{\square})$

d $4.308 = (4 \times \square) + (3 \times \frac{\square}{\square}) + (8 \times \frac{\square}{\square})$

e $6.93 = (6 \times \square) + (9 \times \frac{\square}{\square}) + (3 \times \frac{\square}{\square})$

f $2.015 = (2 \times \square) + (1 \times \frac{\square}{\square}) + (5 \times \frac{\square}{\square})$

6 How would you change the first number to the second number? (For example, to change 1.26 to 1.36, add 1 tenth (0.1).)

a	74.23 to 74.33	**b**	9.36 to 9.16	**c**	42.8 to 44.8
d	3.08 to 3.05	**e**	6.79 to 6.99	**f**	23.4 to 23.7
g	5.16 to 5.12	**h**	8.01 to 7.99	**i**	32.87 to 32.9
j	9.615 to 9.618	**k**	63.17 to 65.17	**l**	4.526 to 4.326

Add and subtract $\frac{1}{10}$, $\frac{1}{100}$ and $\frac{1}{1000}$

1 Write the number that is $\frac{1}{10}$ more.

a	3.64	**b**	12.7	**c**	4.05	**d**	58.9	**e**	12.16	**f**	7.98
g	14.35	**h**	7.01	**i**	5.95	**j**	36.4	**k**	82.36	**l**	4.078
m	2.362	**n**	3.81	**o**	26.2	**p**	4.589	**q**	6.07	**r**	31.92
s	5.99	**t**	17.9	**u**	2.095	**v**	40.712	**w**	3.917	**x**	76.8

2 Write the number that is $\frac{1}{10}$ less.

a	3.8	**b**	7.05	**c**	28	**d**	14.1	**e**	6.72	**f**	51.6
g	8.129	**h**	3.09	**i**	47	**j**	6.152	**k**	4.35	**l**	19.02
m	5.033	**n**	6.14	**o**	13.1	**p**	52.638	**q**	7.81	**r**	6.901
s	8	**t**	3.01	**u**	5.916	**v**	8.03	**w**	16.177	**x**	4.98

3 Write the number that is $\frac{1}{100}$ more.

a	4.86	**b**	3.152	**c**	17.09	**d**	6.395	**e**	27	**f**	5.003
g	2.56	**h**	87.2	**i**	3.96	**j**	9.198	**k**	41.063	**l**	58
m	1.03	**n**	51.76	**o**	2.083	**p**	8.7	**q**	16.59	**r**	70.8
s	5.99	**t**	4.703	**u**	15.7	**v**	33.41	**w**	18	**x**	25.16

4 Write the number that is $\frac{1}{100}$ less.

a	7.83	**b**	5.96	**c**	15.81	**d**	3.67	**e**	39.8	**f**	63.702
g	19.04	**h**	25.001	**i**	36	**j**	5.72	**k**	19.3	**l**	65
m	9.015	**n**	11.6	**o**	20.08	**p**	8.712	**q**	6.073	**r**	12.8
s	49	**t**	5.02	**u**	6.54	**v**	8.137	**w**	3.69	**x**	17.5

5 Write the number that is $\frac{1}{1000}$ more.

a	1.637	**b**	8.045	**c**	2.61	**d**	15.087	**e**	23	**f**	5.57
g	45.09	**h**	32.8	**i**	55.099	**j**	7.4	**k**	29.219	**l**	13.6
m	2.972	**n**	35.009	**o**	85.88	**p**	21.06	**q**	45.089	**r**	3.12
s	5.8	**t**	64	**u**	37.6	**v**	1.409	**w**	81.2	**x**	3.679

6 Write the number that is $\frac{1}{1000}$ less.

a	5.813	**b**	2.655	**c**	17.841	**d**	17	**e**	22.011	**f**	9.82
g	33.094	**h**	57.81	**i**	14.09	**j**	3.8	**k**	48.7	**l**	23.5
m	11.07	**n**	61.11	**o**	32.04	**p**	5.621	**q**	60	**r**	9.5
s	28.6	**t**	3.001	**u**	67.12	**v**	83.8	**w**	7.64	**x**	52

Relate fractions, decimals and percentages

Remember

Fractions with a denominator of 10, 100 or 1000 can be converted to decimals.

For example,

$\frac{7}{10} = 0.7$, $\frac{23}{100} = 0.23$ and $\frac{375}{1000} = 0.375$

1 Convert these fractions to decimals.

a	$\frac{3}{10}$	b	$\frac{9}{10}$	c	$\frac{63}{100}$	d	$\frac{19}{100}$
e	$\frac{3}{100}$	f	$\frac{27}{100}$	g	$\frac{654}{1000}$	h	$\frac{123}{1000}$
i	$\frac{65}{1000}$	j	$\frac{37}{100}$	k	$\frac{1}{10}$	l	$\frac{5}{100}$
m	$\frac{71}{100}$	n	$\frac{9}{1000}$	o	$\frac{6}{10}$	p	$\frac{40}{100}$

2 Convert these mixed numbers to decimals.

a	$1\frac{6}{10}$	b	$7\frac{9}{100}$	c	$3\frac{45}{100}$	d	$17\frac{125}{1000}$	e	$10\frac{41}{100}$
f	$25\frac{75}{100}$	g	$30\frac{4}{100}$	h	$9\frac{87}{1000}$	i	$6\frac{14}{100}$	j	$47\frac{3}{10}$
k	$52\frac{11}{1000}$	l	$8\frac{16}{100}$	m	$2\frac{29}{100}$	n	$31\frac{9}{10}$	o	$4\frac{6}{100}$

3 Write these decimals as fractions in their simplest form.

a	0.8	b	6.75	c	1.317	d	0.087	e	8.005	f	2.19
g	5.03	h	17.16	i	24.052	j	31.8	k	12.06	l	7.509
m	1.023	n	0.67	o	3.411	p	9.12	q	18.75	r	73.6

4 Write these percentages as fractions in their simplest form.

a	25%	b	75%	c	50%	d	7%
e	3%	f	37%	g	78%	h	15%
i	61%	j	10%	k	20%	l	90%
m	39%	n	5%	o	84%	p	19%
q	45%	r	11%	s	9%	t	91%

Remember

Percent means 'for every 100'. Common and decimal fractions can be written as percentages. The symbol for percent is %.

5 Copy and complete the grid.

	Common fraction	Decimal	Percentage
a	$\frac{57}{100}$		
c		0.39	
e			9%
g	$\frac{3}{10}$		

	Common fraction	Decimal	Percentage
b			21%
d		0.7	
f	$\frac{86}{100}$		
h		0.02	

Order and compare fractions, decimals and percentages

1 Which is larger?

a	0.85 or 0.8	b	0.07 or 0.7	c	0.38 or 0.4	d	1.05 or 1.5
e	2.9 or 2.09	f	5.13 or 5.2	g	0.96 or 0.9	h	1.83 or 1.85
i	2.45 or 2.5	j	0.58 or 0.5	k	6.25 or 6.2	l	3.15 or 3.2
m	1.67 or 1.675	n	5.08 or 5.078	o	4.7 or 4.667	p	2.348 or 2.5
q	8.06 or 8.062	r	7.101 or 7.11	s	2.395 or 2.4	t	3.05 or 3.047

2 Write the largest decimal in each group.

a	1.56	1.5	1.65	b	5.7	5.07	5.72	c	9.08	9.86	9.8
d	7.03	7.3	7.05	e	3.52	3.6	3.51	f	2.75	2.15	2.07
g	14	13.83	14.06	h	9.25	9.02	9.3	i	6.3	6.03	6.13
j	8.07	8.7	8.08	k	4.67	4.65	4.06	l	0.16	0.06	0.6
m	1.235	1.2	1.24	n	3.64	3.6	3.638	o	2.801	2.81	2.811

3 Record the smallest fraction, decimal or percentage in each group.

a	0.21	25%	$\frac{2}{10}$	b	63%	0.6	$\frac{7}{10}$	c	$\frac{9}{100}$	10%	0.9
d	35%	$\frac{3}{10}$	0.36	e	0.55	5%	$\frac{5}{10}$	f	$\frac{75}{100}$	0.8	79%
g	$\frac{1}{4}$	20%	0.3	h	80%	$\frac{3}{4}$	0.72	i	51%	0.49	$\frac{1}{2}$
j	$\frac{42}{100}$	40%	0.45	k	0.03	$\frac{3}{10}$	29%	l	96%	0.9	$\frac{95}{100}$
m	0.875	87%	$\frac{88}{100}$	n	$\frac{9}{10}$	93%	0.927	o	0.025	$\frac{2}{100}$	3%

4 Write each group of fractions in order from smallest to largest.

a	65%	0.6	$\frac{61}{100}$	58%	b	81%	0.75	0.08	$\frac{8}{10}$
c	0.48	$\frac{1}{2}$	49%	$\frac{4}{10}$	d	$\frac{2}{10}$	19%	0.02	0.25
e	33%	0.3	$\frac{25}{100}$	31%	f	$\frac{75}{100}$	0.72	7%	0.7
g	0.08	$\frac{8}{10}$	0.75	85%	h	79%	0.7	$\frac{75}{100}$	72%
i	$\frac{35}{100}$	0.3	3%	0.315	j	21%	2.1	$\frac{2}{100}$	0.225
k	1.5	$\frac{10}{100}$	15%	0.015	l	0.5	53%	$\frac{51}{100}$	0.053

Round off decimals

1 Round these weights to the nearest whole kilogram (kg).

a	7.8 kg	**b**	9.2 kg	**c**	1.7 kg	**d**	0.9 kg	**e**	5.4 kg
f	11.6 kg	**g**	3.1 kg	**h**	12.5 kg	**i**	15.9 kg	**j**	10.3 kg
k	8.7 kg	**l**	14.3 kg	**m**	2.75 kg	**n**	6.15 kg	**o**	0.87 kg
p	3.92 kg	**q**	1.39 kg	**r**	4.24 kg	**s**	1.345 kg	**t**	7.033 kg

2 Round these distances to the nearest whole kilometre (km).

a	9.2 km	**b**	6.4 km	**c**	12.9 km	**d**	23.4 km	**e**	15.8 km
f	21.6 km	**g**	19.3 km	**h**	45.1 km	**i**	30.5 km	**j**	26.7 km
k	35.4 km	**l**	28.2 km	**m**	8.25 km	**n**	16.75 km	**o**	27.08 km
p	11.36 km	**q**	17.88 km	**r**	32.53 km	**s**	46.081 km	**t**	18.575 km

3 Round these money amounts to the nearest whole kina (K).

a	K4.55	**b**	K7.89	**c**	K3.35	**d**	K6.08	**e**	K5.86	**f**	K4.19
g	K3.72	**h**	K2.45	**i**	K4.59	**j**	K5.17	**k**	K9.60	**l**	K11.25
m	K15.68	**n**	K10.05	**o**	K18.10	**p**	K16.95	**q**	K8.70	**r**	K7.50

4 Round these capacities to the nearest tenth of a litre (L).

a	4.18 L	**b**	7.56 L	**c**	1.35 L	**d**	0.89 L	**e**	10.65 L	**f**	15.83 L
g	6.07 L	**h**	8.27 L	**i**	12.41 L	**j**	9.33 L	**k**	5.86 L	**l**	2.72 L
m	11.392 L	**n**	17.855 L	**o**	8.076 L	**p**	19.115 L	**q**	7.043 L	**r**	4.903 L

5 Round each time to the nearest second (s) and the nearest tenth of a second (s).

a	10.86 s	**b**	18.93 s	**c**	9.25 s	**d**	11.78 s	**e**	10.99 s	**f**	8.47 s
g	12.34 s	**h**	11.55 s	**i**	8.76 s	**j**	9.03 s	**k**	10.46 s	**l**	12.89 s
m	5.91 s	**n**	15.63 s	**o**	25.17 s	**p**	18.06 s	**q**	24.72 s	**r**	21.29 s

6 Copy and complete this chart by rounding these decimal numbers to the nearest whole number, tenth and hundredth.

	Number	Nearest whole	Nearest $\frac{1}{10}$	Nearest $\frac{1}{100}$
a	2.546			
c	5.081			
e	0.627			
g	9.402			
i	7.354			

	Number	Nearest whole	Nearest $\frac{1}{10}$	Nearest $\frac{1}{100}$
b	1.753			
d	8.245			
f	3.778			
h	6.189			
j	4.885			

Add and subtract decimals

Help Box

Add or subtract decimals with equal numbers of places:

2.76 + 5.45

$$\begin{array}{r} 2.76 \\ +\,5.45 \\ \hline 8.21 \end{array}$$

Add or subtract decimals with unequal numbers of places:

37.31 − 9.5

$$\begin{array}{r} 37.31 \\ -\,9.50 \\ \hline 27.81 \end{array}$$

1 Solve these decimal additions and subtractions.

a 5.8 + 7.6 = ____ b 9.7 + 6.3 = ____ c 8.9 + 5.8 = ____ d 6.4 + 8.5 = ____ e 15.8 + 23.6 = ____ f 25.7 + 36.5 = ____

g 9.7 − 7.5 = ____ h 8.4 − 5.7 = ____ i 16.2 − 12.5 = ____ j 22.3 − 15.6 = ____ k 19.6 − 12.8 = ____ l 32.5 − 24.7 = ____

m 5.77 + 8.69 = ____ n 9.56 + 6.09 = ____ o 25.18 + 18.67 = ____ p 14.35 + 29.85 = ____ q 7.095 + 9.673 = ____ r 8.137 + 6.585 = ____

Remember

When adding and subtracting decimals, keep the decimal points under one another.

2 Solve these decimal additions and subtractions.

a 6.3 + 17.8 = ____ b 27.9 + 36.57 = ____ c 18.09 + 7.637 = ____ d 34.06 + 16.9 = ____ e 4.837 + 67.37 = ____ f 19.56 + 7.852 = ____

g 47.91 − 6.87 = ____ h 51.25 − 8.6 = ____ i 9.054 − 7.76 = ____ j 62.4 − 55.08 = ____ k 37.21 − 28.675 = ____ l 21.443 − 8.78 = ____

Multiply decimals

1 Estimate by rounding to decide where you would place the decimal point in each answer.

- a 6.3×5 = 315
- b 5.7×3 = 171
- c 2.8×6 = 168
- d 7.4×4 = 296
- e 23.8×2 = 476
- f 5.16×4 = 2064
- g 3.89×3 = 1167
- h 2.75×5 = 1375
- i 5.63×7 = 3941
- j 4.58×4 = 1832

Remember

When multiplying decimals, calculate as a normal multiplication and then estimate by rounding to help you place the decimal point in the answer.

2 Solve these multiplications.

- a 24.72×8
- b 32.09×6
- c 183.5×5
- d 156.7×3
- e 27.4×7
- f 3.067×4
- g 7.635×9
- h 8.183×5
- i 28.516×4
- j 37.904×8
- k 521.6×2
- l 8.994×6
- m 4.38×7
- n 76.38×9
- o 9.156×5
- p 145.8×3
- q 17.405×8
- r 8.824×4
- s 589.6×6
- t 41.79×5
- u 3.669×3
- v 156.07×7
- w 29.6×9
- x 4.075×8

3 Calculate the total price of each set of items on the shopping list.

	Item	Number required	Price for one	Total price
a	Boxes of nails	6	K10.55	
b	Hammers	3	K24.50	
c	Tubes of glue	8	K3.49	
d	Buckets	5	K6.95	
e	Rolls of wire	4	K12.65	
f	Sandpaper sheets	12	K0.87	

Divide decimals

1 Estimate by rounding to decide where you would place the decimal point in each answer.

Remember

When dividing decimals, calculate as a normal division and then estimate by rounding to help you place the decimal point in the answer.

a $\begin{array}{r} 34 \\ 2\overline{)6.8} \end{array}$ **b** $\begin{array}{r} 32 \\ 3\overline{)9.6} \end{array}$ **c** $\begin{array}{r} 62 \\ 4\overline{)24.8} \end{array}$ **d** $\begin{array}{r} 31 \\ 6\overline{)18.6} \end{array}$

e $\begin{array}{r} 41 \\ 5\overline{)20.5} \end{array}$ **f** $\begin{array}{r} 221 \\ 4\overline{)8.84} \end{array}$ **g** $\begin{array}{r} 232 \\ 3\overline{)6.96} \end{array}$ **h** $\begin{array}{r} 71 \\ 5\overline{)35.5} \end{array}$

i $\begin{array}{r} 101 \\ 6\overline{)60.6} \end{array}$ **j** $\begin{array}{r} 1322 \\ 3\overline{)39.66} \end{array}$ **k** $\begin{array}{r} 2413 \\ 2\overline{)48.26} \end{array}$ **l** $\begin{array}{r} 311 \\ 7\overline{)21.77} \end{array}$

2 Solve these divisions.

a $2\overline{)2.8}$ **b** $4\overline{)8.4}$ **c** $3\overline{)9.3}$ **d** $5\overline{)5.5}$

e $7\overline{)14.7}$ **f** $2\overline{)16.4}$ **g** $6\overline{)12.6}$ **h** $4\overline{)16.8}$

i $3\overline{)24.63}$ **j** $5\overline{)25.55}$ **k** $8\overline{)32.16}$ **l** $3\overline{)15.96}$

m $6\overline{)7.26}$ **n** $4\overline{)5.68}$ **o** $7\overline{)9.17}$ **p** $5\overline{)7.55}$

q $8\overline{)9.68}$ **r** $3\overline{)8.19}$ **s** $2\overline{)7.46}$ **t** $4\overline{)9.24}$

u $5\overline{)657.5}$ **v** $6\overline{)852.6}$ **w** $4\overline{)712.8}$ **x** $2\overline{)535.2}$

y $7\overline{)93.94}$ **z** $3\overline{)52.62}$

Decimal word problems

Decide which process to use and then solve each word problem.

1 Tau weighs 57.2 kg and Suari weighs 39.7 kg. How much more than Suari does Tau weigh?

2 One metre of fabric costs K5.60. How much do the following lengths cost?

a 3 metres **b** 7 metres **c** 6 metres **d** 8 metres

3 A tourist spent K590.50 on airfares and K312.70 on accommodation.

a How much is this all together?

b How much more did the tourist spend on airfares than accommodation?

4 Ben has travelled 156.25 kilometres. He still has another 87.5 kilometres to travel. What is the total distance Ben will travel?

5 The total price of three books is K24.84. Each book is the same price. How much is each book?

6 A truck carried six boxes of books. Each box weighed 45.3 kg. What was the total weight of the boxes?

7 Ruth received K27.45 change at the trade store. She gave the owner K50. How much did she spend?

8 A water tank contains 1750.65 litres. If 885.9 litres were used, how much is left in the tank?

9 A plane made three flights. The first flight was 3567.8 kilometres, the second flight was 1874.5 kilometres and the third flight was 985.7 kilometres. What was the total distance for the three flights?

10 A bucket holds 7.85 litres. How much will eight of these buckets hold in total?

11 Our journey is a distance of 240.3 kilometres. We have already travelled 156.6 kilometres. How much further do we have to travel?

12 A roll of chicken wire 24.8 metres long was used to completely enclose a square garden. How long is each side of the garden?

13 Naomi has K1763.45 in her bank account. She needs to withdraw K327.50 and K186.35 to pay bills. How much will be left in her account after she makes the withdrawals?

Assessment Number and Application

Multiple choice test for Number and place value (5.1.1)

1 Which number matches the words *thirty-eight thousand and sixty-four*?

a 38 640 **b** 38 064 **c** 3864 **d** 38 604

2 Which number is 1000 more than *forty-nine thousand and twenty-seven*?

a 40 027 **b** 59 027 **c** 50 027 **d** 49 127

3 Which number is more than 81 516?

a 80 900 **b** 81 500 **c** 81 520 **d** 81 000

4 Which number is between 16 850 and 17 025?

a 16 700 **b** 17 010 **c** 17 100 **d** 16 810

5 Which number is largest?

a 9063 **b** 9306 **c** 9603 **d** 9630

6 Which number is smallest?

a 72 835 **b** 72 385 **c** 72 538 **d** 72 358

7 Which number is closest numerically to 43 056?

a 43 040 **b** 43 000 **c** 43 100 **d** 43 500

8 What number does 400 + 10 + 7000 + 20 000 + 6 make?

a 27 461 **b** 72 641 **c** 27 416 **d** 27 406

9 In which number is 8 worth the most?

a 98 470 **b** 82 533 **c** 95 518 **d** 92 803

10 In which number is 3 worth the least?

a 26 131 **b** 23 417 **c** 35 006 **d** 15 388

11 Which number is 10 more than 11 890?

a 11 899 **b** 11 990 **c** 11 880 **d** 11 900

12 Which number is 100 more than 37 958?

a 38 058 **b** 37 058 **c** 38 098 **d** 38 158

13 What has been added to 51 075 to make 51 175?

a 10 **b** 100 **c** 1000 **d** 10 000

14 What has been added to 12 399 to make 12 409?

a 1000 **b** 100 **c** 10 **d** 1

15 In which number is 5 worth 50?

a 85 109 **b** 76 058 **c** 32 517 **d** 28 115

16 K3080 rounded to the nearest K100 is:

a K3800 **b** K3000 **c** K4000 **d** K3100

17 18 516 rounded to the nearest 1000 is:

a 19 000 **b** 18 000 **c** 20 000 **d** 18 500

18 Which one of these numbers is a square number?

a 18 **b** 27 **c** 64 **d** 35

19 Which one of these numbers is a prime number?

a 36 **b** 29 **c** 40 **d** 27

20 Which one of these numbers is cubed?

a 9 **b** 125 **c** 65 **d** 36

Operations (5.1.2)

1 Add each set of numbers in your head and write the answers.

a	17 + 6 + 20 + 4	**b**	13 + 20 + 9 + 7	**c**	19 + 16 + 29 + 14
d	8 + 15 + 12 + 16	**e**	25 + 9 + 25 + 18	**f**	11 + 17 + 19 + 13

2 Add each pair of numbers in your head and write the answers.

a	76 + 29	**b**	99 + 38	**c**	68 + 11	**d**	126 + 39	**e**	53 + 101
f	209 + 98	**g**	149 + 19	**h**	51 + 138	**i**	199 + 119	**j**	121 + 249

3 Subtract each pair of numbers in your head and write the answers.

a	52 – 9	**b**	71 – 19	**c**	63 – 11	**d**	45 – 8	**e**	84 – 7
f	120 – 49	**g**	150 – 19	**h**	130 – 21	**i**	322 – 9	**j**	361 – 39

4 Multiply these numbers in your head and write the answers.

a	6 × 7	**b**	4 × 8	**c**	9 × 6	**d**	8 × 12	**e**	8 × 7
f	12 × 6	**g**	9 × 8	**h**	7 × 5	**i**	8 × 8	**j**	7 × 7

5 Divide these numbers in your head and write the answers.

a	63 ÷ 9	**b**	40 ÷ 8	**c**	84 ÷ 12	**d**	28 ÷ 7	**e**	60 ÷ 12
f	48 ÷ 8	**g**	21 ÷ 7	**h**	81 ÷ 9	**i**	45 ÷ 9	**j**	108 ÷ 12

6 Solve these multiplications and divisions.

a	80 × 4	**b**	6 × 30	**c**	7 × 50	**d**	60 × 6	**e**	80 × 7
f	140 ÷ 2	**g**	320 ÷ 8	**h**	270 ÷ 3	**i**	480 ÷ 8	**j**	360 ÷ 9
k	30 × 20	**l**	50 × 30	**m**	40 × 50	**n**	30 × 40	**o**	60 × 30
p	720 ÷ 80	**q**	810 ÷ 90	**r**	490 ÷ 70	**s**	630 ÷ 90	**t**	420 ÷ 60

7 Solve these additions and subtractions.

a	50 + 80	**b**	70 + 40	**c**	60 + 50	**d**	90 + 80	**e**	30 + 90
f	100 – 30	**g**	80 – 50	**h**	120 – 70	**i**	150 – 90	**j**	230 – 60
k	170 + 60	**l**	140 + 70	**m**	190 + 80	**n**	230 + 70	**o**	250 + 90
p	130 – 50	**q**	210 – 40	**r**	340 – 60	**s**	200 – 80	**t**	310 – 70

8 Do these calculations in your head and write the answers.

a	315 – 99	**b**	257 + 101	**c**	199 × 4	**d**	3300 ÷ 33	**e**	649 + 250	**f**	6 × 89
g	4000 ÷ 800	**h**	523 – 102	**i**	7 × 401	**j**	433 – 199	**k**	199 + 278	**l**	800 ÷ 50

9 Find:

a the difference between 53 and 25
b the total of 36, 29 and 35
c the product of 10 and 8
d the quotient of 40 and 4
e the sum of 60 and 125
f 72 divided by 6
g 9 times 19
h 92 minus 28

10 Set out these additions and then solve each one.

a 4623 + 3376
b 2364 + 4525
c 5773 + 2862
d 17 069 + 12 388
e 28 517 + 36 965
f 19 455 + 32 708
g 20 625 + 7816 + 29 813
h 9455 + 3607 + 15 839
i 7238 + 53 631 + 9865
j 6751 + 847 + 31 098
k 6347 + 44 390 + 668
l 35 468 + 19 803 + 5574

11 Set out these subtractions and then solve each one.

a 8759 – 5236
b 7688 – 2465
c 9546 – 6032
d 6142 – 3851
e 8206 – 5989
f 7335 – 4875
g 33 524 – 19 618
h 41 042 – 23 587
i 28 315 – 17 832
j 52 327 – 8865
k 65 403 – 9658
l 47 324 – 6477

12 Solve these multiplications.

a 2316×3
b 1745×6
c 3082×5
d 2758×4
e 4607×7
f $13\,760 \times 5$
g $24\,258 \times 8$
h $11\,968 \times 3$
i $32\,607 \times 6$
j $25\,943 \times 4$

13 Solve these long multiplications.

a 4382×22
b 1879×35
c 3622×17
d 5066×24
e 2059×38

14 Solve these divisions. Write any remainders as fractions.

a $5321 \div 7$
b $4637 \div 5$
c $7341 \div 4$
d $5972 \div 6$
e $4669 \div 8$
f $6107 \div 3$
g $8042 \div 6$
h $2594 \div 7$
i $3778 \div 4$
j $1683 \div 5$
k $16\,728 \div 8$
l $23\,049 \div 3$
m $21\,566 \div 5$
n $19\,703 \div 7$
o $35\,824 \div 4$

15 Solve these long divisions. Write any remainders as fractions.

a $17\overline{)6885}$
b $25\overline{)4793}$
c $31\overline{)6908}$
d $22\overline{)5729}$
e $18\overline{)5377}$
f $36\overline{)7345}$
g $29\overline{)8408}$
h $21\overline{)3827}$
i $23\overline{)36\,142}$
j $19\overline{)53\,015}$
k $32\overline{)81\,992}$
l $27\overline{)48\,596}$

Fractions (5.1.3)

1 Find the fraction of each group.

a	$\frac{1}{2}$ of 24	b	$\frac{1}{4}$ of 16	c	$\frac{1}{3}$ of 33	d	$\frac{1}{5}$ of 40
e	$\frac{1}{8}$ of 24	f	$\frac{1}{10}$ of 70	g	$\frac{1}{7}$ of 21	h	$\frac{1}{6}$ of 48
i	$\frac{1}{9}$ of 18	j	$\frac{1}{4}$ of 36	k	$\frac{3}{4}$ of 28	l	$\frac{2}{3}$ of 15
m	$\frac{4}{5}$ of 30	n	$\frac{7}{10}$ of 50	o	$\frac{5}{6}$ of 18	p	$\frac{3}{7}$ of 35
q	$\frac{5}{8}$ of 32	r	$\frac{3}{5}$ of 45	s	$\frac{3}{8}$ of 64	t	$\frac{5}{7}$ of 21

2 Solve these problems.

a A PMV carries 30 people. $\frac{3}{5}$ of the people are adults. How many are not adults?

b Of a group of 18 animals, $\frac{2}{3}$ are pigs. How many are not pigs?

c There were 32 fish caught. $\frac{7}{8}$ of the fish were large. How many were not large?

d Naomi peeled $\frac{3}{4}$ of a bag of 24 potatoes. How many did she peel?

e Wesley has 64 marbles. $\frac{3}{4}$ of the marbles are patterned. How many are not patterned?

3 Convert these improper fractions to mixed numbers.

a	$\frac{17}{7}$	b	$\frac{20}{9}$	c	$\frac{15}{4}$	d	$\frac{25}{6}$	e	$\frac{31}{5}$	f	$\frac{19}{3}$
g	$\frac{14}{3}$	h	$\frac{35}{8}$	i	$\frac{20}{3}$	j	$\frac{37}{4}$	k	$\frac{26}{7}$	l	$\frac{23}{4}$
m	$\frac{21}{8}$	n	$\frac{43}{5}$	o	$\frac{22}{7}$	p	$\frac{53}{10}$	q	$\frac{29}{9}$	r	$\frac{13}{6}$

4 Convert these mixed numbers to improper fractions.

a	$2\frac{7}{8}$	b	$1\frac{5}{6}$	c	$3\frac{5}{7}$	d	$5\frac{5}{8}$	e	$3\frac{3}{4}$	f	$4\frac{2}{3}$
g	$1\frac{4}{5}$	h	$5\frac{9}{10}$	i	$4\frac{3}{5}$	j	$6\frac{1}{2}$	k	$2\frac{3}{8}$	l	$7\frac{2}{5}$
m	$3\frac{3}{8}$	n	$1\frac{5}{9}$	o	$5\frac{2}{3}$	p	$2\frac{8}{9}$	q	$9\frac{1}{4}$	r	$4\frac{3}{7}$

5 Write equivalent fractions.

a $\frac{2}{3} = \frac{\square}{12}$

b $\frac{16}{18} = \frac{\square}{9}$

c $\frac{3}{5} = \frac{\square}{25}$

d $\frac{10}{25} = \frac{\square}{5}$

e $\frac{4}{7} = \frac{\square}{14}$

f $\frac{18}{30} = \frac{\square}{5}$

g $\frac{5}{8} = \frac{\square}{32}$

h $\frac{18}{24} = \frac{\square}{4}$

i $\frac{3}{8} = \frac{\square}{32}$

j $\frac{30}{50} = \frac{\square}{5}$

k $\frac{7}{9} = \frac{\square}{54}$

l $\frac{21}{28} = \frac{\square}{4}$

6 Add or subtract these fractions.

a $\frac{5}{9} + \frac{6}{9}$

b $\frac{13}{15} - \frac{11}{15}$

c $\frac{4}{5} + \frac{4}{5}$

d $\frac{11}{12} - \frac{6}{12}$

e $\frac{7}{12} + \frac{2}{3}$

f $\frac{9}{10} - \frac{3}{5}$

g $\frac{7}{8} + \frac{3}{4}$

h $\frac{11}{15} - \frac{3}{5}$

i $\frac{2}{3} + \frac{5}{9}$

j $\frac{11}{12} - \frac{3}{4}$

k $\frac{5}{6} + \frac{7}{12}$

l $\frac{5}{8} - \frac{1}{4}$

7 Add or subtract these mixed numbers.

a $1\frac{5}{8} + 4\frac{3}{4}$

b $4\frac{9}{10} - 1\frac{4}{5}$

c $3\frac{5}{6} + 2\frac{2}{3}$

d $5\frac{1}{2} - 2\frac{1}{6}$

e $2\frac{1}{3} + 5\frac{5}{9}$

f $4\frac{7}{12} - 1\frac{1}{6}$

g $1\frac{7}{8} + 6\frac{1}{2}$

h $6\frac{2}{3} - 2\frac{5}{12}$

i $3\frac{4}{5} + 2\frac{7}{10}$

j $5\frac{3}{4} - 2\frac{3}{8}$

k $2\frac{7}{12} + 3\frac{1}{4}$

l $4\frac{5}{7} - 1\frac{5}{14}$

Decimals (5.1.4)

1 What is the value of the underlined digit?

a $2\underline{7}.63$ **b** $162.\underline{5}$ **c** $3.0\underline{8}6$ **d** $2.\underline{7}5$ **e** $0.18\underline{9}$ **f** $1.\underline{2}83$

g $12.4\underline{4}$ **h** $\underline{5}.802$ **i** $\underline{9}3.6$ **j** $34.\underline{4}1$ **k** $6.3\underline{9}$ **l** $7.58\underline{4}$

2 What numbers are these?

a 4 tenths, 2 ones, 8 hundredths, 5 tens

b 7 thousandths, 9 ones, 4 hundredths, 1 tenth

c 3 ones, 5 hundredths, 6 tens, 2 thousandths

d 8 hundredths, 2 tenths, 6 ones, 9 thousandths

3 Write the number that is $\frac{1}{10}$ more.

a 0.63 **b** 15.8 **c** 27.652 **d** 8.06 **e** 4.95 **f** 12.9

4 Write the number that is $\frac{1}{100}$ less.

a 5.83 **b** 22.6 **c** 36.074 **d** 9.11 **e** 24 **f** 0.105

5 Convert these fractions and mixed numbers to decimals.

a $\frac{7}{10}$ **b** $\frac{39}{100}$ **c** $\frac{3}{100}$ **d** $\frac{147}{1000}$ **e** $\frac{78}{1000}$

f $\frac{9}{1000}$ **g** $2\frac{5}{10}$ **h** $8\frac{27}{100}$ **i** $12\frac{80}{100}$ **j** $6\frac{1}{100}$

k $5\frac{83}{1000}$ **l** $10\frac{69}{1000}$ **m** $2\frac{19}{100}$ **n** $15\frac{8}{1000}$ **o** $26\frac{30}{100}$

6 Write these decimals as fractions in their simplest form.

a 0.6 **b** 3.47 **c** 1.09 **d** 6.751 **e** 2.037 **f** 11.4

7 Write these percentages as fractions in their simplest form.

a 50% **b** 30% **c** 9% **d** 23% **e** 6% **f** 55%

8 Write these percentages as decimals.

a 75% **b** 3% **c** 56% **d** 81% **e** 8% **f** 49%

9 Write each group of fractions, decimals and percentages in order from largest to smallest.

a 79%, 0.079, $\frac{7}{100}$, 0.8

b 0.02, $\frac{2}{10}$, 25%, 0.23

c 0.33, 35%, $\frac{30}{100}$, $\frac{3}{100}$

d 58%, 0.058, 5.8, $\frac{5}{10}$

e 87%, 0.8, $\frac{87}{1000}$, 0.08

f 0.15, 1.5, $\frac{15}{1000}$, 10%

10 Round these numbers to the nearest tenth.

a 7.36 **b** 4.07 **c** 2.513 **d** 9.081 **e** 5.25 **f** 8.705

11 Round these numbers to the nearest hundredth.

a 1.367 **b** 0.818 **c** 5.033 **d** 8.214 **e** 6.045 **f** 2.402

12 Solve these additions and subtractions.

a 7.5 + 8.6 **b** 8.9 − 5.7 **c** 16.3 + 8.8 **d** 9.2 − 6.5 **e** 32.8 + 19.7 **f** 24.3 − 16.8

g 42.6 + 9.54 **h** 32.5 − 18.42 **i** 7.867 + 5.08 **j** 5.77 − 2.583 **k** 17.09 + 508.3 **l** 62.15 − 35.8

13 Solve these multiplications.

a 5.8 × 4 **b** 8.7 × 7 **c** 3.8 × 5 **d** 7.63 × 3 **e** 2.57 × 6 **f** 23.08 × 8

g 31.07 × 5 **h** 6.742 × 4 **i** 9.058 × 8 **j** 514.9 × 9 **k** 237.6 × 7 **l** 41.64 × 6

14 Solve these divisions.

a 3)6.9 **b** 7)21.7 **c** 4)12.8 **d** 5)30.5 **e** 6)8.46 **f** 5)9.55

g 8)13.76 **h** 7)17.78 **i** 4)923.6 **j** 9)196.2 **k** 3)581.4 **l** 6)279.6

15 Solve these word problems.

a Daniel had K750 in his bank account. Over two days he withdrew K96.50 and K116.85. How much is still in his account?

b Nahau made 7 dresses. She used 1.76 metres of fabric for each dress. How much fabric did she use altogether?

c Elsie sold 47.63 kg of potatoes the first day, 64.12 kg the second day and 55.8 kg the third day. How many kilograms of potatoes did Elsie sell over the three days?

5.2.1 Estimate, measure and solve problems using standard units of length

Estimate, measure and record length

1 Draw a chart like the one below with 18 rows. Estimate the length of each line below the chart, then measure each line and record its length. The first one has been done for you.

	Estimate	Length in mm	Length in cm and mm	Decimal
a	60 mm	65 mm	6 cm 5 mm	6.5 cm

a ______ b ______

c ______

d ______ e ______

f ______ g ______

h ______ i ______

j ______

k ______ l ______

m ______

n ______ o ______

p ______

q ______ r ______

2 Draw lines that you estimate to be these lengths and then measure them to see how close you were.

a 50 mm **b** 7 cm 5 mm **c** 30 mm **d** 1 cm
e 2.3 cm **f** 10.2 cm **g** 6.4 cm **h** 8.1 cm

Convert length units

1 Write these metre (m) lengths as centimetres (cm).

a	2 m	**b**	5 m	**c**	1 m	**d**	4 m	**e**	10 m
f	12 m	**g**	$3\frac{1}{2}$ m	**h**	$6\frac{1}{2}$ m	**i**	$1\frac{1}{2}$ m	**j**	$9\frac{1}{2}$ m

2 Write these centimetre (cm) lengths as metres (m).

a	700 cm	**b**	300 cm	**c**	800 cm	**d**	1100 cm	**e**	600 cm
f	750 cm	**g**	250 cm	**h**	450 cm	**i**	1250 cm	**j**	1550 cm

3 Write these centimetre (cm) lengths as millimetres (mm).

a	8 cm	**b**	5 cm	**c**	11 cm	**d**	3 cm	**e**	14 cm
f	$2\frac{1}{2}$ cm	**g**	$8\frac{1}{2}$ cm	**h**	$10\frac{1}{2}$ cm	**i**	$4\frac{1}{2}$ cm	**j**	$7\frac{1}{2}$ cm

4 Write these millimetre (mm) lengths as centimetres (cm).

a	50 mm	**b**	10 mm	**c**	100 mm	**d**	70 mm	**e**	120 mm
f	65 mm	**g**	115 mm	**h**	25 mm	**i**	95 mm	**j**	15 mm

5 Write these kilometre (km) lengths as metres (m).

a	2 km	**b**	5 km	**c**	9 km	**d**	4 km	**e**	10 km
f	$5\frac{1}{2}$ km	**g**	$\frac{1}{2}$ km	**h**	$11\frac{1}{2}$ km	**i**	$6\frac{1}{2}$ km	**j**	$3\frac{1}{2}$ km

6 Write these metre (m) lengths as kilometres (km).

a	1000 m	**b**	6000 m	**c**	3000 m	**d**	8000 m	**e**	13 000 m
f	2500 m	**g**	8500 m	**h**	16 500 m	**i**	1500 m	**j**	10 500 m

7 Write these lengths using metres and centimetres. (For example, 187 cm = 1 m 87 cm)

a	123 cm	**b**	247 cm	**c**	108 cm	**d**	361 cm	**e**	296 cm
f	1314 cm	**g**	2559 cm	**h**	1282 cm	**i**	3405 cm	**j**	5117 cm

8 Write these lengths using centimetres and millimetres. (For example, 54 mm = 5 cm 4 mm)

a	89 mm	**b**	16 mm	**c**	63 mm	**d**	125 mm	**e**	271 mm
f	103 mm	**g**	334 mm	**h**	218 mm	**i**	566 mm	**j**	485 mm

9 Write these lengths using kilometres and metres. (For example, 2150 m = 2 km 150 m)

a	1700 m	**b**	1345 m	**c**	3678 m	**d**	2097 m	**e**	1820 m
f	14 562 m	**g**	10 204 m	**h**	47 035 m	**i**	32 712 m	**j**	26 955 m

Measure perimeter

Measure the perimeter of these shapes in centimetres.

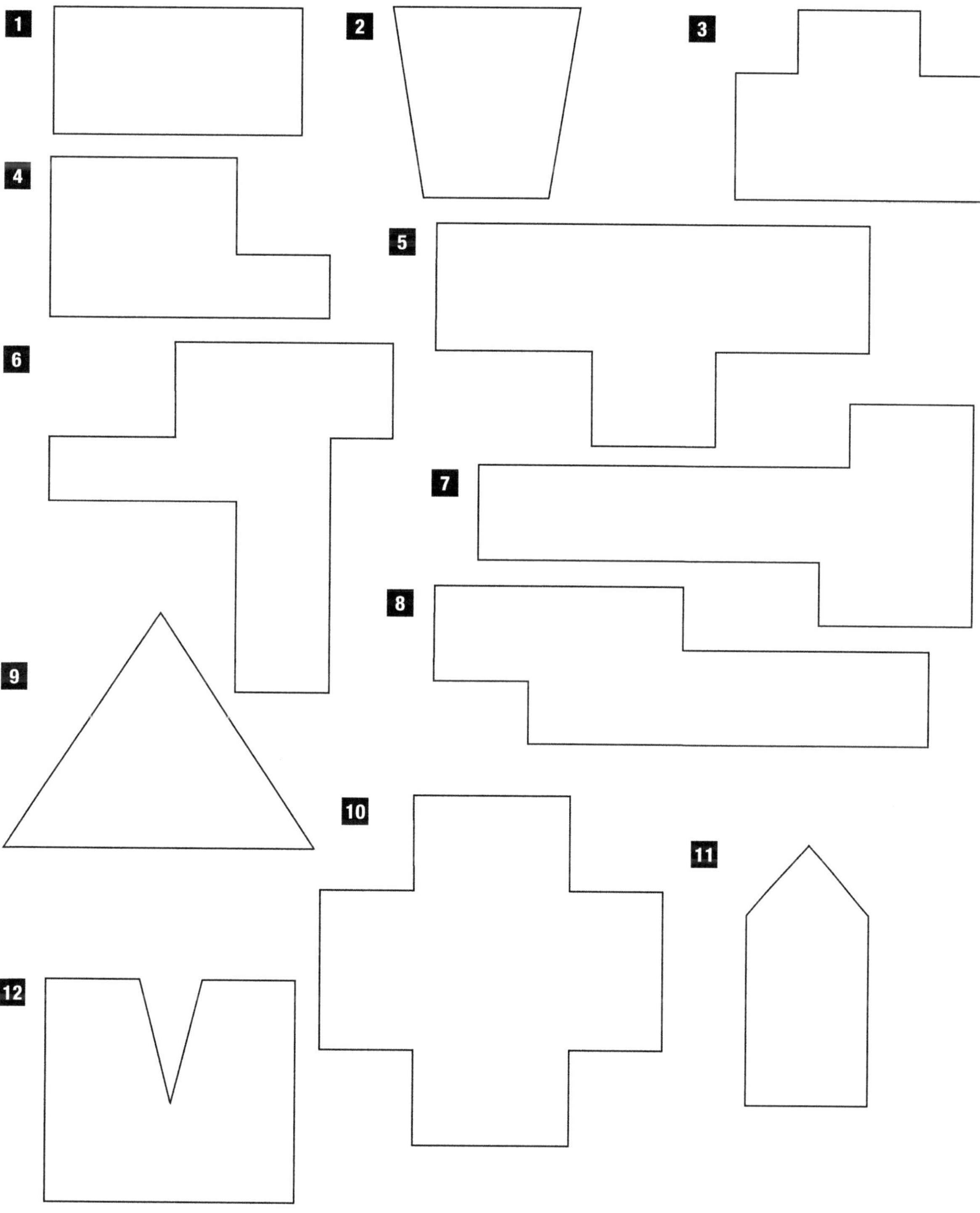

5.2.2 Use appropriate metric units to measure and calculate area

Find area by counting square centimetres

Count the centimetre squares to find the area in square centimetres (cm^2) of each shape.

1

2

3

4

5

6

7

8

9

10

11

Use a grid to find area

Draw a chart like the one below with 12 rows and complete it by recording the length, width and area of each rectangle on the page.

	Length (cm)	Width (cm)	Area (cm^2)
1			
2			
3			

1

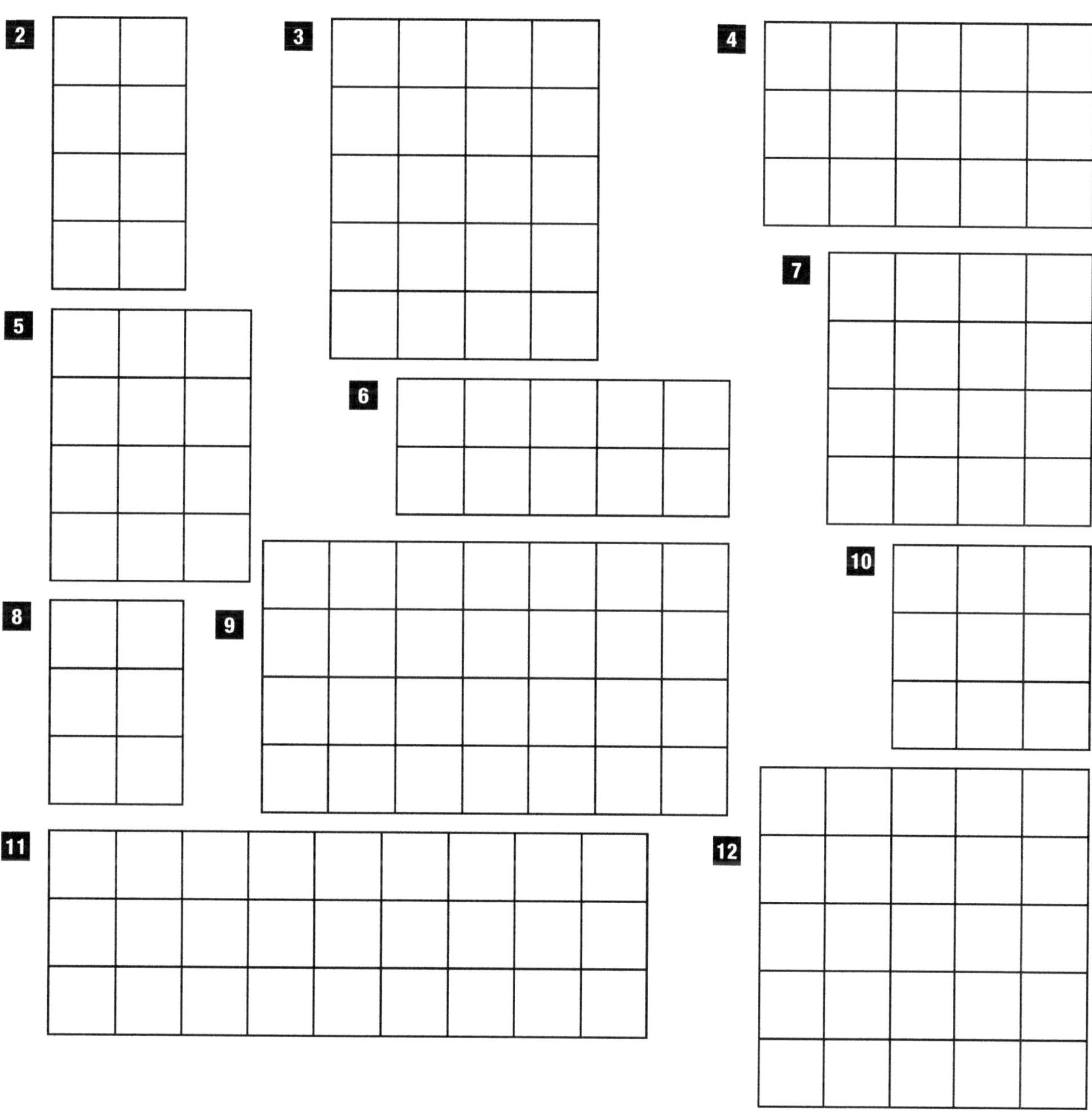

Calculate area of rectangles

Calculate the area of these rectangles.

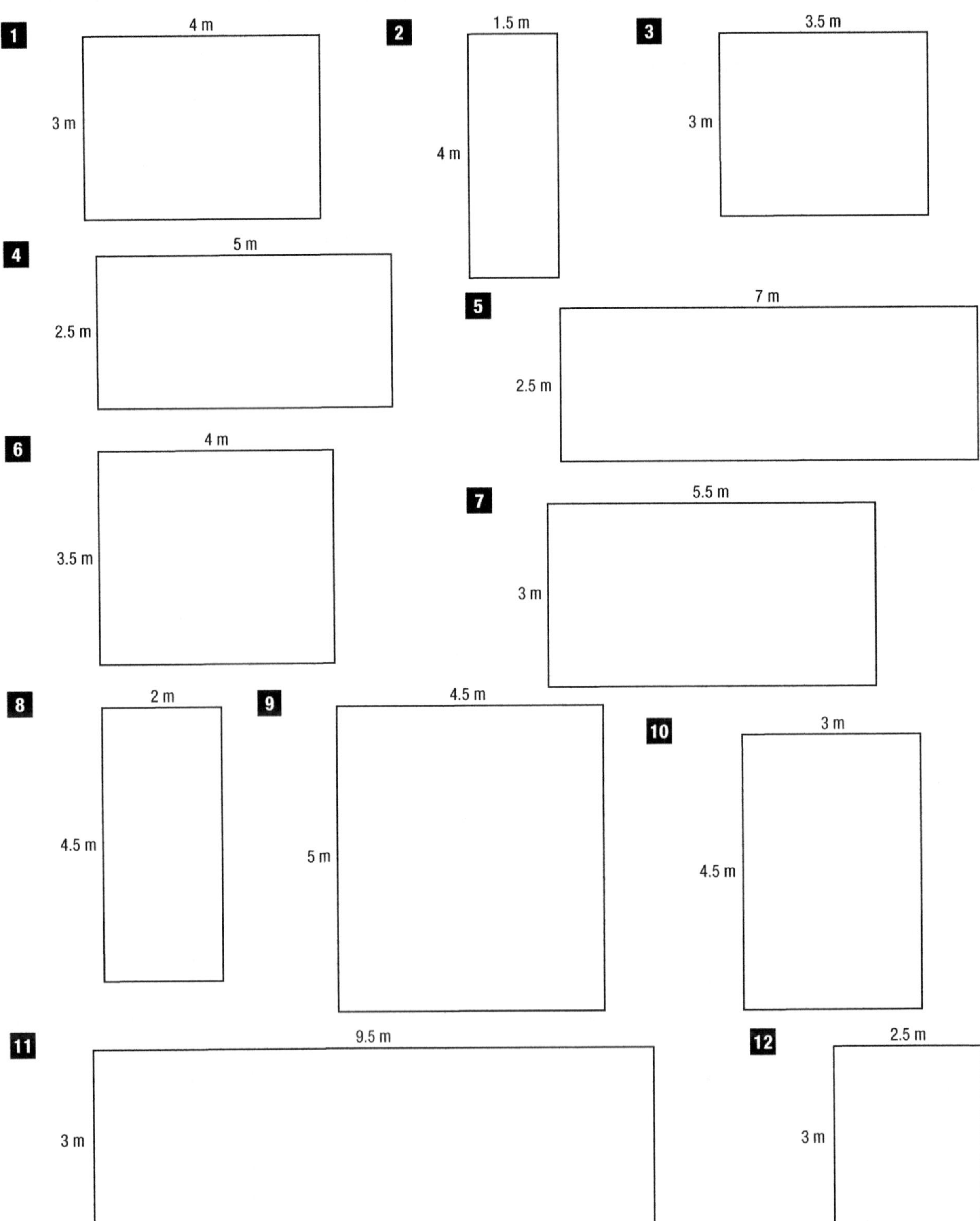

5.2.3 Estimate, measure and solve problems using standard units of volume and capacity

Convert capacity and volume units

1 Write these quantities using litres (L) and millilitres (mL). (For example, 2563 mL = 2 L 563 mL)

a	1540 mL	**b**	2735 mL	**c**	5061 mL	**d**	1007 mL
e	3604 mL	**f**	7010 mL	**g**	4218 mL	**h**	3720 mL
i	2906 mL	**j**	1700 mL	**k**	6194 mL	**l**	2733 mL
m	4372 mL	**n**	5995 mL	**o**	2036 mL	**p**	1877 mL

2 Write these quantities as millilitres (mL).

a	8 L	**b**	3 L	**c**	$5\frac{1}{2}$ L	**d**	$2\frac{1}{2}$ L	**e**	6 L 70 mL
f	8 L 510 mL	**g**	1 L 15 mL	**h**	3 L 407 mL	**i**	$\frac{1}{4}$ L	**j**	$4\frac{1}{4}$ L
k	9.5 L	**l**	0.5 L	**m**	7.326 L	**n**	2.75 L	**o**	5.6 L
p	0.532 L	**q**	4.9 L	**r**	3.15 L	**s**	1.7 L	**t**	8.25 L

3 Use decimal notation to write these quantities as litres (L).

a	3500 mL	**b**	5250 mL	**c**	2750 mL	**d**	500 mL	**e**	2765 mL	**f**	4860 mL
g	1485 mL	**h**	3038 mL	**i**	250 mL	**j**	900 mL	**k**	5002 mL	**l**	4800 mL

4 Find the total of these quantities.

a	0.65 L + 1250 mL + 5 L	**b**	375 mL + $2\frac{1}{2}$ L + 4.8 L	**c**	$1\frac{1}{4}$ L + 3.6 L + 2500 mL
d	2.43 L + 580 mL + 0.7 L	**e**	1140 mL + $3\frac{1}{2}$ L + 1.35 L	**f**	$5\frac{3}{4}$ L + 0.4 L + 1638 mL
g	7 L + 2630 mL + 3.475 L	**h**	1.7 L + 2.45 L + $\frac{3}{10}$ L	**i**	$\frac{1}{10}$ L + 6.02 L + 885 mL
j	$\frac{4}{5}$ L + 1227 mL + 1.139 L				

5 Write these cubic centimetres (cm^3) as cubic metres (m^3).

a	200 000 cm^3	**b**	500 000 cm^3	**c**	800 000 cm^3
d	300 000 cm^3	**e**	150 000 cm^3	**f**	450 000 cm^3
g	1 000 000 cm^3	**h**	650 000 cm^3	**i**	325 000 cm^3
j	975 000 cm^3	**k**	225 000 cm^3	**l**	460 000 cm^3
m	130 000 cm^3	**n**	570 000 cm^3	**o**	810 000 cm^3
p	380 000 cm^3				

Remember

Cubic centimetres (cm^3) and cubic metres (m^3) are used to measure volume.

100 000 cm^3 = 1 m^3

6 Write these cubic metres (m^3) as cubic centimetres (cm^3).

a	5 m^3	**b**	9 m^3	**c**	6 m^3	**d**	2.5 m^3	**e**	$7\frac{1}{2}$ m^3
f	1.25 m^3	**g**	$4\frac{1}{4}$ m^3	**h**	5.75 m^3	**i**	0.5 m^3	**j**	$10\frac{3}{4}$ m^3
k	2.6 m^3	**l**	5.9 m^3	**m**	$11\frac{1}{2}$ m^3	**n**	1.3 m^3	**o**	$12\frac{1}{4}$ m^3
p	3.4 m^3	**q**	$2\frac{3}{10}$ m^3	**r**	$3\frac{1}{5}$ m^3	**s**	$1\frac{7}{10}$ m^3	**t**	$5\frac{3}{5}$ m^3

Find the volume of irregular solids

1 Each cube is one cubic metre. Find the volume of each model by counting the cubes.

A

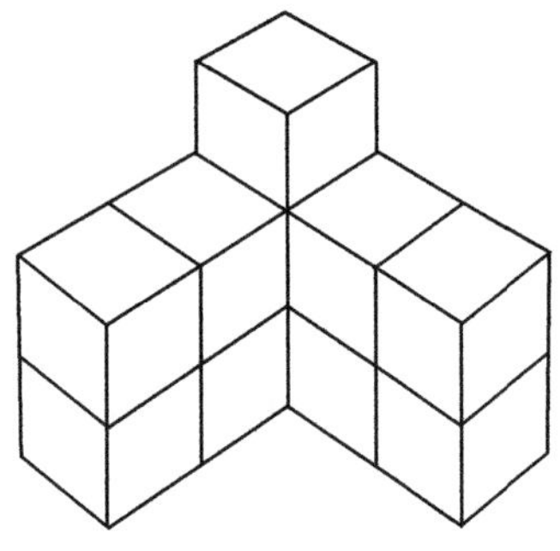

B

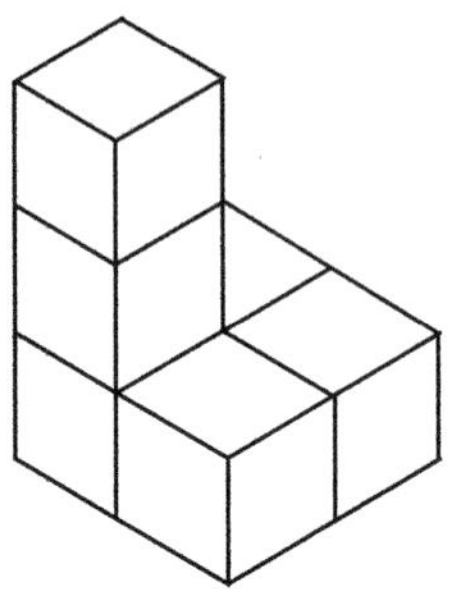

C

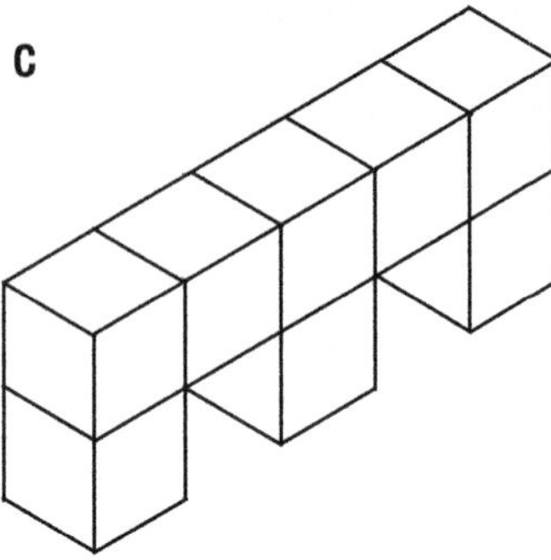

D

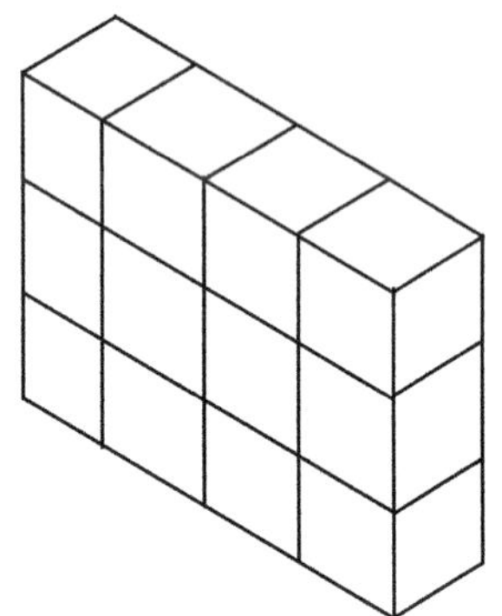

E

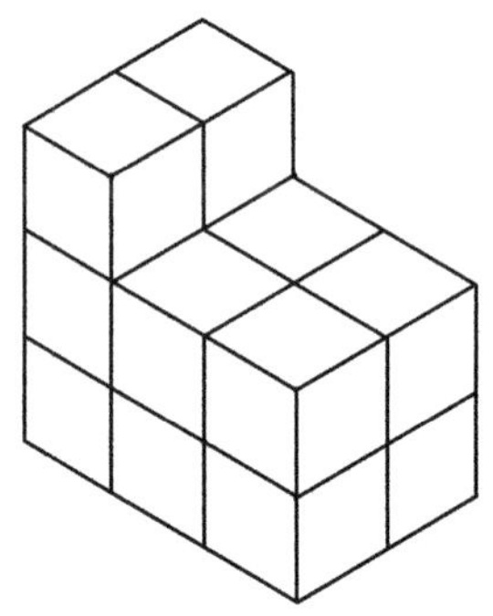

F

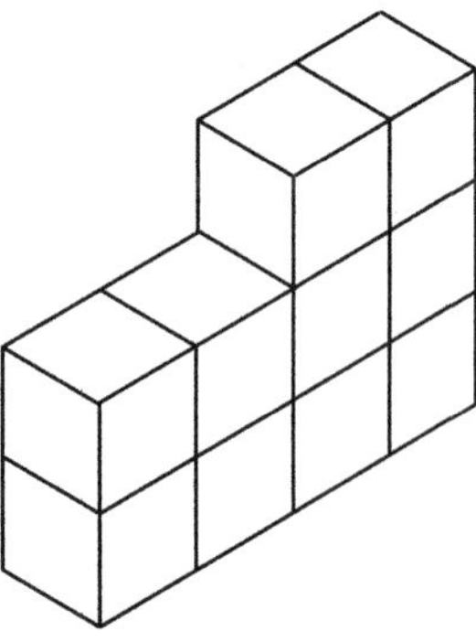

G

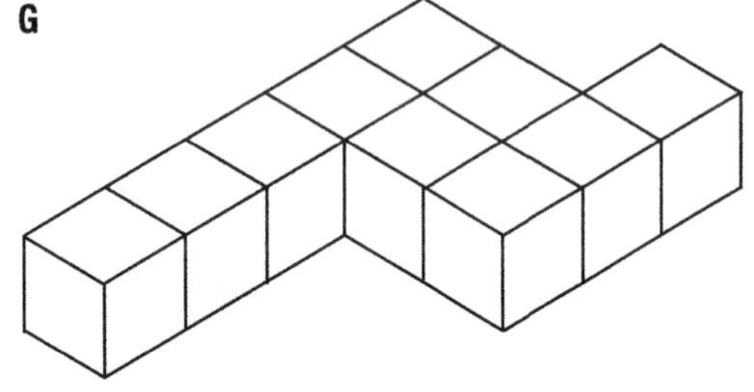

H

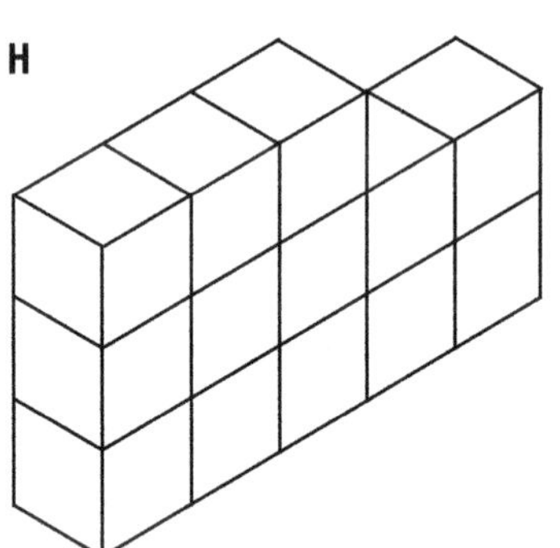

I

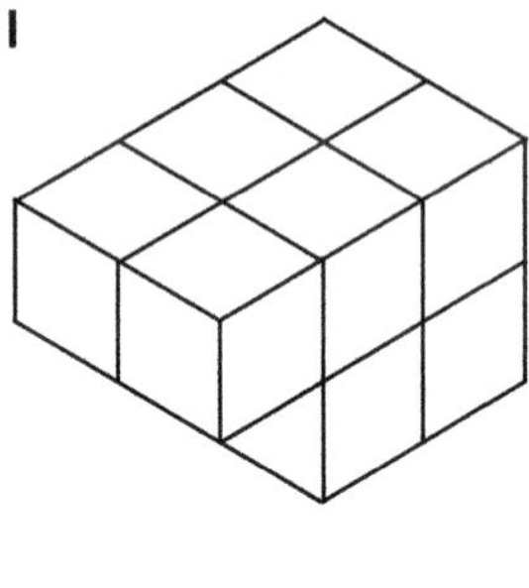

2 Which model has the greatest volume?

3 Which model has the smallest volume?

4 What is the difference in volume between

a **H** and **B**?   b **D** and **G**?   c **A** and **F**?   d **C** and **H**?

5 What is the total volume of

a **F** and **E**?   b **D** and **A**?   c **I** and **B**?   d **G** and **C**?

Find the volume of rectangular prisms

Copy and complete the chart to find the volume of each rectangular prism.

	Prism	Length in centimetres	Width in centimetres	Height in centimetres	Volume
1					
2					
3					
4					
5					
6					
7					
8					

To find the volume of a rectangular prism, multiply its length by its width by its height.

Length = 5 cm
Width = 3 cm
Height = 2 cm

Volume = Length × Width × Height
$= 5 \times 3 \times 2 = 30\ cm^3$

5.2.4 Estimate, measure and solve problems using standard units of weight

Compare and order weight

1 Write the weights that are more than $\frac{1}{2}$ kilogram (kg).

675 g 1.2 kg 350 g 0.25 kg 0.86 kg
0.615 kg 1010 g 1.7 kg 4.1 kg 0.38 kg
725 g 0.15 kg 510 g 375 g 1250 g

2 Write the weights that are less than $\frac{1}{4}$ kilogram (kg).

0.29 kg 0.09 kg 476 g 240 g 1.95 kg
100 g 0.075 kg 1.08 kg 0.05 kg 106 g
480 g 0.136 kg 87 g 1.6 kg 0.175 kg

3 Write the weights that are between $\frac{1}{4}$ kilogram and $\frac{3}{4}$ kilogram.

0.5 kg 615 g 450 g 0.312 kg 0.756 kg
508 g 0.068 kg 185 g 7.5 kg 0.74 kg
806 g 1.95 kg 1.15 kg 0.607 kg 0.083 kg

4 Write the weights that are between $\frac{3}{4}$ kilogram and $1\frac{1}{4}$ kilograms.

0.78 kg 1.585 kg 1.67 kg 596 g 1300 g
1240 g 0.98 kg 2150 g 1.02 kg 0.68 kg
1.5 kg 1.065 kg 0.95 kg 350 g 1.23 kg

5 Write the weights that are less than $\frac{3}{4}$ tonne (t).

	650 kg		0.875 t		970 kg		1.5 t		0.35 t
0.4 t		1200 kg		740 kg		0.25 t		2.75 t	
	0.7 t		0.912 t		400 kg		7500 kg		347 kg

6 Write the weights that are more than $\frac{4}{5}$ tonne (t).

	700 kg		1350 kg		0.65 t		820 kg		1.25 t
0.9 t		667 kg		0.75 t		1275 kg		0.85 t	
	815 kg		764 kg		0.91 t		1.45 t		1000 kg

7 Write each group of weights in order from lightest to heaviest.

a	6.5 kg	1690 g	$6\frac{1}{4}$ kg	6150 g
b	0.3 t	275 kg	0.035 t	305 kg
c	7.55 t	7500 kg	7.03 t	0.75 t
d	1.45 kg	1600 g	1300 g	$1\frac{1}{2}$ kg
e	534 g	0.5 kg	0.053 kg	0.53 kg
f	4100 kg	4.01 t	4.15 t	4050 kg
g	806 kg	8.6 t	8.06 t	0.87 t
h	2.3 kg	230 g	2030 g	2.13 kg
i	493 g	0.49 kg	4.9 kg	0.049 kg
j	1.6 t	1550 kg	1.065 t	1.16 t
k	9 t	0.9 t	8900 kg	0.95 t
l	1300 g	1.34 kg	1.03 kg	1285 g
m	5.2 kg	5270 g	5.025 kg	$5\frac{1}{4}$ kg
n	0.75 t	0.075 t	715 kg	0.7 t
o	$2\frac{1}{2}$ t	2550 kg	2.15 t	255 kg
p	8100 g	8.17 kg	8.09 kg	$8\frac{1}{4}$ kg
q	1785 g	1.7 kg	$1\frac{3}{4}$ kg	1.075 kg
r	6.7 t	6070 kg	6.17 t	6.075 t

Convert between metric units of weight

1 How many grams (g)?

a	$\frac{1}{2}$ kg	**b**	$\frac{1}{5}$ kg	**c**	$2\frac{1}{2}$ kg	**d**	$\frac{1}{4}$ kg	**e**	$\frac{3}{4}$ kg	**f**	$\frac{1}{10}$ kg
g	$\frac{7}{10}$ kg	**h**	$3\frac{1}{5}$ kg	**i**	$\frac{1}{8}$ kg	**j**	$2\frac{4}{5}$ kg	**k**	$1\frac{3}{4}$ kg	**l**	$4\frac{1}{4}$ kg
m	$4\frac{3}{4}$ kg	**n**	$2\frac{2}{5}$ kg	**o**	$8\frac{1}{2}$ kg	**p**	$1\frac{1}{4}$ kg	**q**	$\frac{4}{5}$ kg	**r**	$\frac{3}{8}$ kg

2 How many kilograms (kg)?

a	$\frac{3}{4}$ t	**b**	$3\frac{1}{2}$ t	**c**	$2\frac{1}{4}$ t	**d**	$\frac{7}{8}$ t	**e**	$1\frac{9}{10}$ t	**f**	$10\frac{1}{2}$ t
g	$3\frac{3}{8}$ t	**h**	$\frac{3}{5}$ t	**i**	$1\frac{5}{8}$ t	**j**	$5\frac{7}{8}$ t	**k**	$6\frac{3}{10}$ t	**l**	$11\frac{1}{4}$ t
m	$12\frac{3}{8}$ t	**n**	$10\frac{1}{5}$ t	**o**	$9\frac{7}{8}$ t	**p**	$11\frac{4}{5}$ t	**q**	$15\frac{1}{2}$ t	**r**	$12\frac{3}{4}$ t

3 Convert these grams to kilograms and fractions of kilograms (kg).

a	4500 g	**b**	7500 g	**c**	2250 g	**d**	8250 g	**e**	2750 g
f	6750 g	**g**	5100 g	**h**	4125 g	**i**	7900 g	**j**	6125 g

4 Write these amounts as kilograms (kg).

a	2.5 t	**b**	7.5 t	**c**	4.25 t	**d**	9.75 t	**e**	0.25 t	**f**	1.75 t
g	5.6 t	**h**	8.3 t	**i**	0.9 t	**j**	10.2 t	**k**	4.8 t	**l**	12.4 t
m	6.125 t	**n**	3.078 t	**o**	0.193 t	**p**	0.672 t	**q**	6.934 t	**r**	5.007 t

5 Write these amounts as decimal fractions of a kilogram (kg).

a	2300 g	**b**	4100 g	**c**	1600 g	**d**	850 g	**e**	2630 g	**f**	1875 g
g	3560 g	**h**	795 g	**i**	4238 g	**j**	3080 g	**k**	2704 g	**l**	4160 g
m	425 g	**n**	1036 g	**o**	8010 g	**p**	7300 g	**q**	239 g	**r**	4129 g

6 Change these kilograms to tonnes and fractions of tonnes (t).

a	1500 kg	**b**	10 500 kg	**c**	6250 kg	**d**	3250 kg	**e**	11 750 kg
f	9750 kg	**g**	12 100 kg	**h**	15 900 kg	**i**	10 200 kg	**j**	9125 kg

7 Write these amounts as grams (g).

a	0.6 kg	**b**	1.4 kg	**c**	6.5 kg	**d**	2.7 kg	**e**	9.1 kg	**f**	3.8 kg
g	11.25 kg	**h**	7.25 kg	**i**	12.75 kg	**j**	20.65 kg	**k**	10.42 kg	**l**	8.17 kg
m	0.156 kg	**n**	1.907 kg	**o**	5.077 kg	**p**	11.04 kg	**q**	15.03 kg	**r**	9.163 kg

5.2.5 Compare and discuss the relationship between units of time and tell time correctly

Convert between time units

1 How many seconds in these minutes?

a 5 minutes b 10 minutes c 4 minutes d $2\frac{1}{2}$ minutes e $8\frac{1}{4}$ minutes

f $3\frac{3}{4}$ minutes g $6\frac{1}{2}$ minutes h $1\frac{1}{3}$ minutes i 12 minutes j 30 minutes

2 How many minutes in these hours?

a 3 hours b 7 hours c $3\frac{1}{2}$ hours d $2\frac{1}{4}$ hours e $5\frac{1}{3}$ hours

f $1\frac{3}{4}$ hours g $10\frac{1}{2}$ hours h $4\frac{1}{5}$ hours i $1\frac{2}{3}$ hours j $\frac{3}{4}$ hour

3 How many hours in these days?

a 2 days b 5 days c 10 days d 7 days e $3\frac{1}{2}$ days

f $8\frac{1}{2}$ days g $6\frac{1}{4}$ days h $1\frac{1}{4}$ days i $4\frac{3}{4}$ days j $9\frac{3}{4}$ days

4 Write these minutes as hours and minutes.

a 185 minutes b 75 minutes c 230 minutes d 310 minutes e 145 minutes

f 90 minutes g 253 minutes h 191 minutes i 372 minutes j 500 minutes

5 Write these hours as days and hours.

a 30 hours b 50 hours c 250 hours d 27 hours e 47 hours

f 75 hours g 33 hours h 55 hours i 25 hours j 100 hours

6 Write these days as weeks and days.

a 10 days b 20 days c 16 days d 40 days e 25 days

f 30 days g 72 days h 50 days i 36 days j 45 days

7 Fill the gaps.

a 4 years = ________ months b $7\frac{1}{2}$ years = ________ months

c 8 decades = ________ years d 15 decades = ________ years

e 6 centuries = ________ years f $2\frac{3}{4}$ centuries = ________ years

Read and record time to five minutes

Write the time from each clock face in analogue and digital form. For example, the first clock shows $\frac{1}{2}$ past 2 or 2:30.

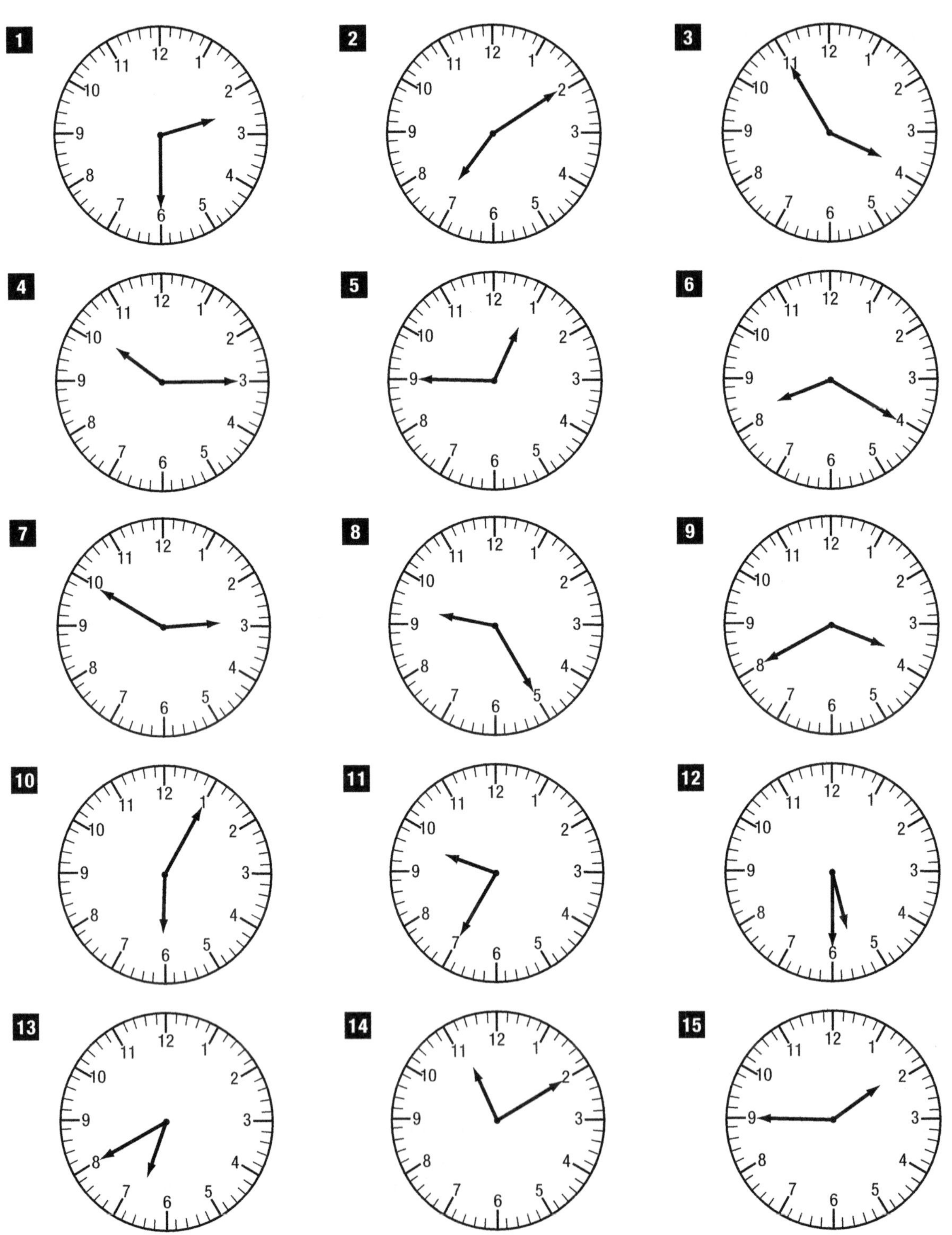

a.m. and p.m. times

1 From the times below, write the ones that are in the morning.

9:20 a.m. 1:30 p.m. 7:30 p.m. 6:45 a.m. 2:50 p.m.
3:50 a.m. 4:15 p.m. 11:20 a.m. 7:30 a.m. 6:15 p.m.
2:10 a.m. 7:05 a.m. 9:25 p.m. 8:15 p.m. 6:30 a.m.

2 Put each group of times in order from earliest to latest.

a	7:30 a.m.	7:15 p.m.	11 a.m.	**b**	1:35 p.m.	1:50 a.m.	2:15 p.m.
c	4:45 p.m.	3:15 p.m.	5:20 p.m.	**d**	6:40 a.m.	5:45 a.m.	6:15 p.m.
e	9:05 a.m.	9:10 p.m.	8:55 p.m.	**f**	2:10 p.m.	2:30 a.m.	2:50 a.m.
g	3:30 p.m.	3:45 p.m.	3:25 p.m.	**h**	8:45 a.m.	8:15 p.m.	8:55 a.m.
i	11:10 p.m.	11 a.m.	11:40 p.m.	**j**	5:25 a.m.	5:15 a.m.	5:25 p.m.

3 Copy and complete the chart below. Use a.m. and p.m. when writing times.

	Time	1 hour later	$\frac{1}{2}$ hour later	$1\frac{1}{2}$ hours earlier	15 minutes earlier
a	11:45 a.m.				
b	10:30 a.m.				
c	1:15 p.m.				
d	12:30 p.m.				
e	1 p.m.				
f	11:15 a.m.				
g	12:50 p.m.				
h	11:35 p.m.				
i	11:10 p.m.				
j	1 a.m.				
k	12:45 a.m.				
l	12 noon				
m	11 p.m.				
n	10:45 p.m.				
o	1:10 a.m.				

Convert between 12-hour and 24-hour clock time

1 Write a 24-hour time from the panel to match each 12-hour time.

a	5:35 p.m.	**b**	9:30 p.m.
c	11:40 p.m.	**d**	6:30 a.m.
e	10:45 a.m.	**f**	1:20 p.m.
g	8:15 a.m.	**h**	4:15 a.m.
i	2:10 p.m.	**j**	11:05 a.m.
k	10:55 p.m.	**l**	4 p.m.
m	7:50 a.m.	**n**	3:25 p.m.
o	10 a.m.	**p**	9:55 a.m.

1105 0815 1735 1525
0955 1410 1600 1320
2340 0630 2255 0415
1045 0750 2130 1000

2 Write these 24-hour clock times as 12-hour clock times. (For example, 1330 = 1:30 p.m.)

a	1415	**b**	1830	**c**	0705	**d**	1040	**e**	0645	**f**	1550
g	0425	**h**	1855	**i**	1935	**j**	2210	**k**	2345	**l**	2015
m	1205	**n**	0045	**o**	0850	**p**	1325	**q**	2115	**r**	1510
s	1640	**t**	1735	**u**	2245	**v**	0205	**w**	1920	**x**	1340

3 Write these 12-hour clock times as 24-hour clock times.

a	4:15 a.m.	**b**	5:50 p.m.	**c**	11:30 p.m.	**d**	1:15 p.m.	**e**	3:20 a.m.
f	7:45 a.m.	**g**	7 p.m.	**h**	9:35 a.m.	**i**	8:30 p.m.	**j**	11 p.m.
k	10:55 p.m.	**l**	6:10 p.m.	**m**	2:25 p.m.	**n**	4:45 p.m.	**o**	5:50 a.m.
p	10 a.m.	**q**	11:10 a.m.	**r**	3:35 p.m.	**s**	9:15 a.m.	**t**	1:05 p.m.
u	2:35 p.m.	**v**	4 p.m.	**w**	10:40 a.m.	**x**	12:10 p.m.	**y**	12:35 a.m.

4 What is the flying time for each plane? (For example, a plane that departs at 1150 and arrives at 1300 has a flying time of 1 hour and 10 minutes.)

	Flight	Depart	Arrive
a	PX 167	1400	1520
b	PX 312	1130	1310
c	PX 045	1545	1700
d	PX 131	1015	1205

	Flight	Depart	Arrive
e	PX 246	1625	1815
f	PX 028	1110	1320
g	PX 255	0950	1420
h	PX 167	1855	2030

	Flight	Depart	Arrive
i	PX 022	1710	1900
j	PX 325	1050	1240
k	PX 183	2140	2355
l	PX 217	2245	0045

5 Write each group of times in order from earliest to latest.

a	1:30 p.m. 2 p.m. 1:20 p.m.	**b**	0720 0800 0745	**c**	2130 8:15 p.m. 2005
d	6:15 a.m. 0640 6:05 a.m.	**e**	1705 1755 5:15 p.m.	**f**	10:40 p.m. 2215 10:20 p.m.
g	9 p.m. 2030 2105	**h**	4:30 p.m. 1610 4 a.m.	**i**	1540 4 p.m. 3:30 a.m.
j	1810 6:05 p.m. 1825	**k**	1:15 p.m. 1345 12:30 p.m.	**l**	1450 2:30 p.m. 1405

Assessment Measurement

1 Fill the gaps.

a 6 m = ________ cm
b $8\frac{1}{2}$ m = ________ cm
c 1150 cm = ________ m
d 4 cm = ________ mm
e $12\frac{1}{2}$ cm = ________ mm
f 150 mm = ________ cm
g 85 mm = ________ cm
h 7 km = ________ m
i $10\frac{1}{2}$ km = ________ m
j 6000 m = ________ km
k 19 500 m = ________ km
l 14 500 m = ________ km

2 Measure the perimeter of these shapes in centimetres.

a
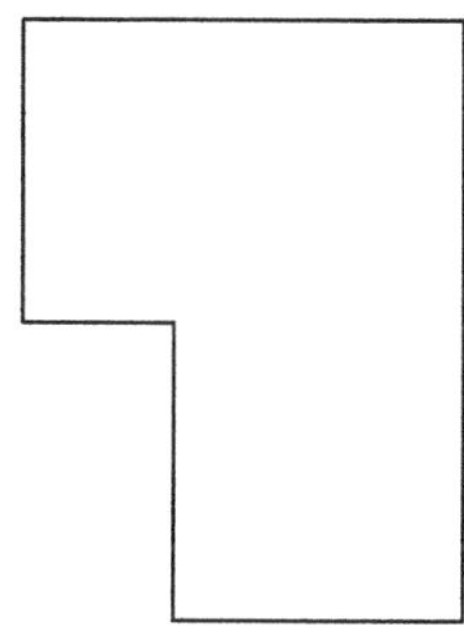

b
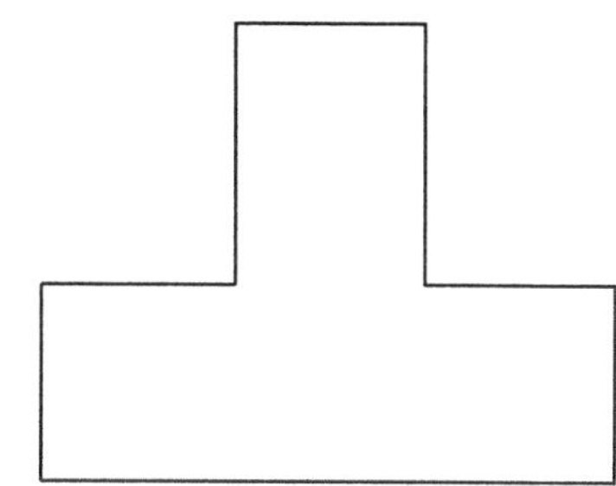

c
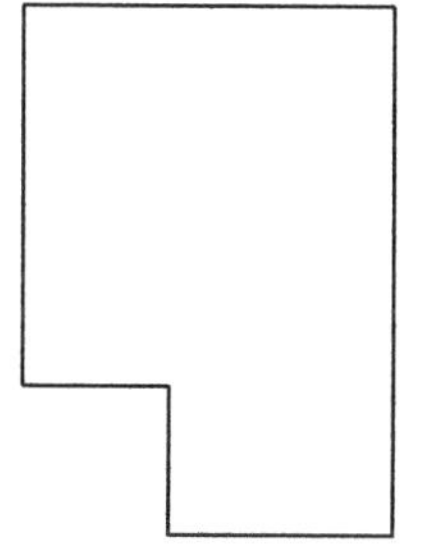

3 Calculate the area of these rectangles. They are not drawn to scale.

a 7 m, 2 m

b 5 m, 3.5 m

c 3 m, 4.5 m

4 Fill the gaps.

a 5 L = ___ mL
b $3\frac{1}{2}$ L = ___ mL
c $\frac{3}{4}$ L = ___ mL
d 8.5 L = ___ mL
e 6.7 L = ___ mL
f 2.15 L = ___ mL
g 1500 mL = ___ L
h 700 mL = ___ L
i 4250 mL = ___ L
j 400 000 cm^3 = ___ m^3
k 950 000 cm^3 = ___ m^3
l 725 000 cm^3 = ________ m^3
m 1 m^3 = ______ cm^3
n $5\frac{1}{2}$ m^3 = ______ cm^3
o 6.75 m^3 = ________ cm^3

5 If each small cube is 1 cm × 1 cm × 1 cm, what is the volume of each rectangular prism?

a
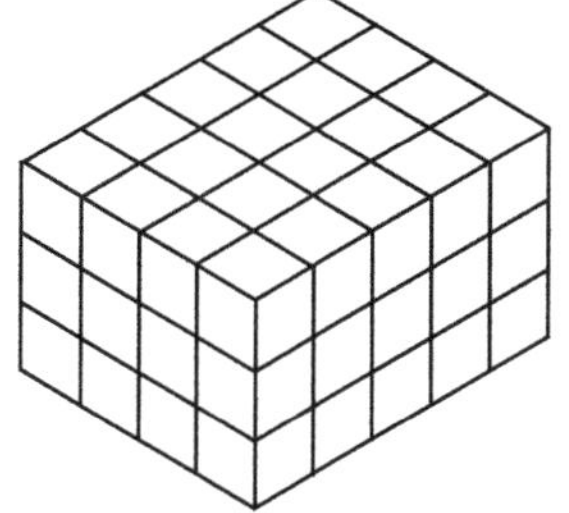

b
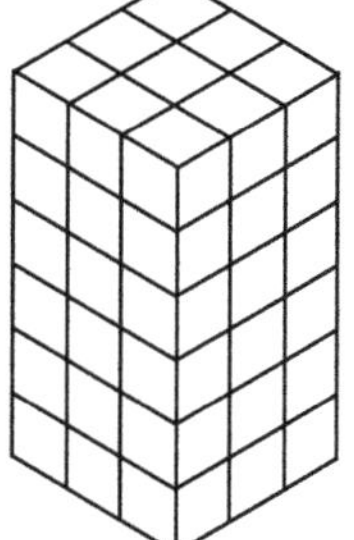

c

6 Order these weights from heaviest to lightest.

a $3\frac{3}{4}$ kg, 3.5 kg, 350 g, 3150 g
b 7.65 kg, 7500 g, $7\frac{1}{4}$ kg, 7.065 kg
c 1.8 kg, 0.85 kg, 8500 g, 1850 g
d $5\frac{1}{2}$ t, 5750 kg, 5.7 t, 5000 kg
e 0.15 t, 0.015 t, 115 kg, 0.1 t
f 2.8 t, 280 kg, 2080 kg, 2.008 t

7 Fill the gaps.

a $9\frac{1}{2}$ minutes = ________ seconds
b $4\frac{1}{4}$ hours = ________ minutes
c 150 minutes = ________ hours
d $6\frac{1}{3}$ days = ________ hours
e $3\frac{3}{4}$ days = ________ hours
f 96 hours = ________ days
g 8 weeks = ________ days
h 84 days = ________ weeks
i 5 years = ________ months
j $3\frac{1}{4}$ years = ________ months
k 144 months = ________ years
l 90 months = ________ years
m $3\frac{1}{2}$ decades = ________ years
n $4\frac{1}{2}$ centuries = ________ years

8 Write the time from each clock in analogue and digital form.

a 
b 
c 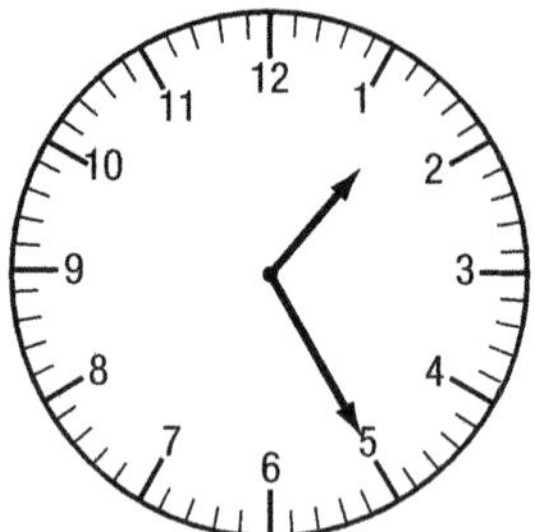

9 What is the time 15 minutes later than these times?

a	11:50 a.m.	b	2:15 p.m.	c	8 p.m.	d	10:20 a.m.	e	5:45 p.m.	f	7:05 a.m.
g	11:55 p.m.	h	3:25 p.m.	i	9:50 a.m.	j	4:10 p.m.	k	5:35 a.m.	l	2:50 a.m.

10 What is the time $\frac{1}{2}$ hour earlier than these times?

a	12:20 p.m.	b	10:05 a.m.	c	2:15 p.m.	d	8:50 a.m.	e	6:10 p.m.	f	1:25 p.m.
g	11:40 a.m.	h	7:55 a.m.	i	6:20 a.m.	j	4:35 p.m.	k	12:05 p.m.	l	9:25 p.m.

11 Write these 24-hour clock times as 12-hour clock times.

a	1315	b	2030	c	0810	d	1145	e	1525	f	2235
g	0505	h	0045	i	1720	j	2340	k	1055	l	1450

12 Write these 12-hour clock times as 24-hour clock times.

a	3:20 a.m.	b	2:45 p.m.	c	10 p.m.	d	11:50 a.m.	e	5:35 p.m.
f	7:30 p.m.	g	9:15 a.m.	h	6:10 a.m.	i	1:55 p.m.	j	9:05 p.m.

Strand	Space and Shape

5.3.1 Describe properties of two and three-dimensional shapes

Describe two-dimensional and three-dimensional shapes

1 Copy and complete the following chart for two-dimensional shapes. The first one has been done for you.

	Shape	Name	Number of sides	Number of vertices (corners)	Number and type of angles	Lines of symmetry
a	△	triangle	3	3	3 acute	3
b	□					
c	⬠					
d	⬡					
e						
f						
g						
h						

2 Copy and complete this chart for three-dimensional shapes. The first one has been done for you.

	Shape	Name	Number of vertices (corners)	Number of faces (flat surfaces)	Number of edges (straight lines)	Curved surface?
a	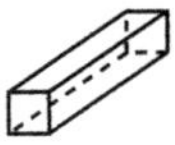	square prism	8	6	12	No
b	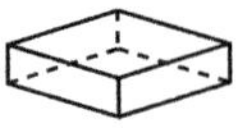					
c						
d						
e						
f						
g	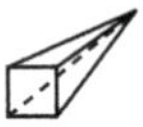					
h						
i						

Match nets to three-dimensional shapes

Copy each shape and draw the net/s that will make the shape.

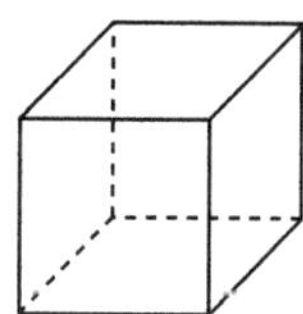

a 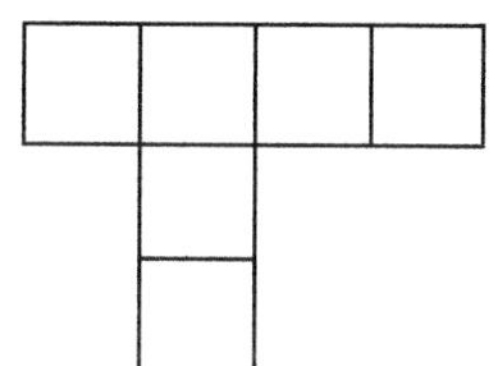

b 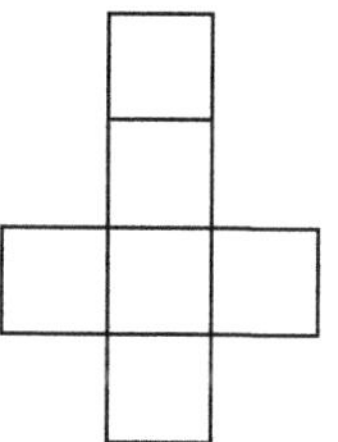

c 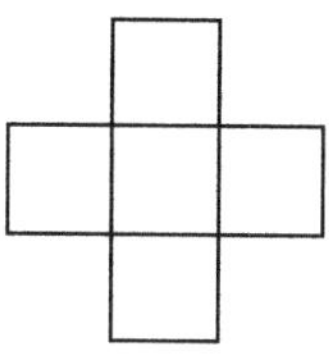

2

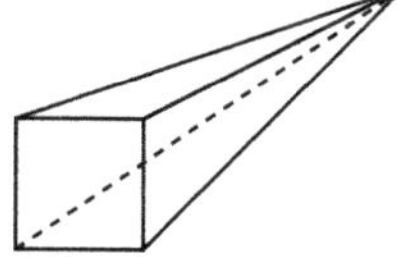

a 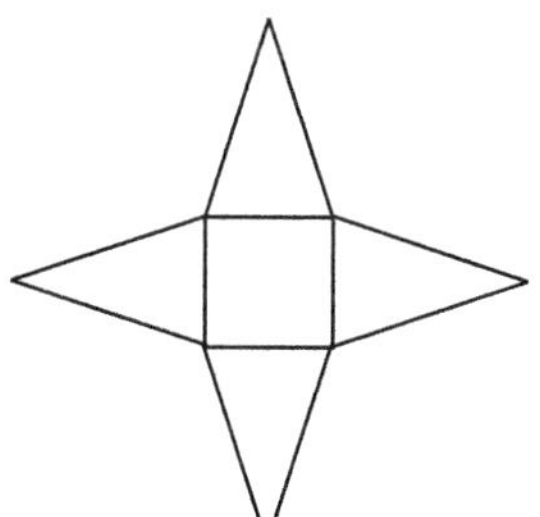

b 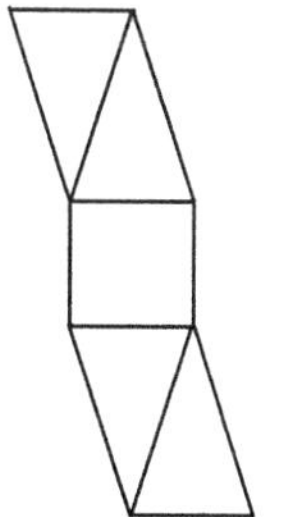

c 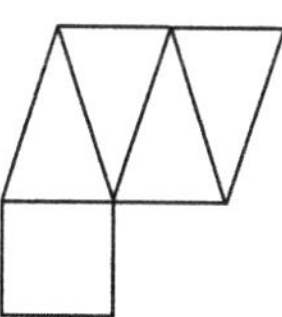

3

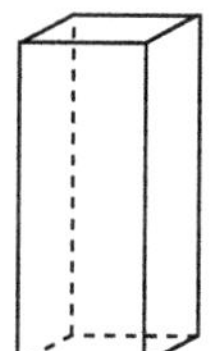

a 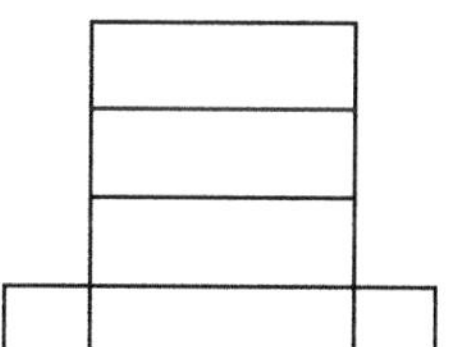

b 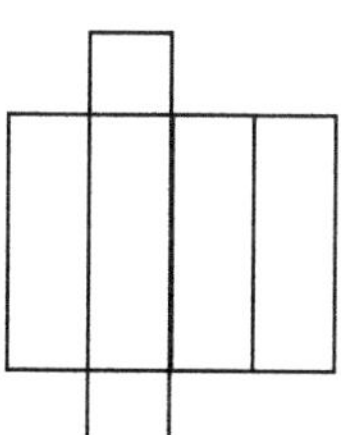

c 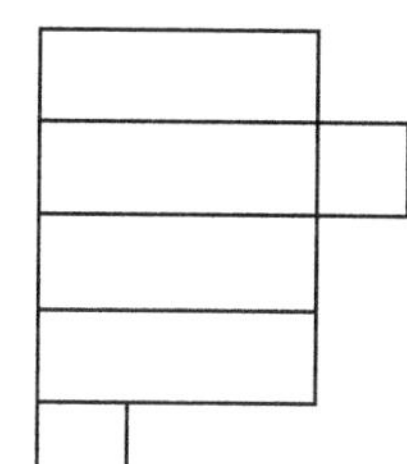

4

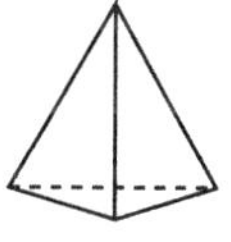

a 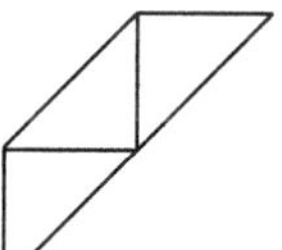

b 

c 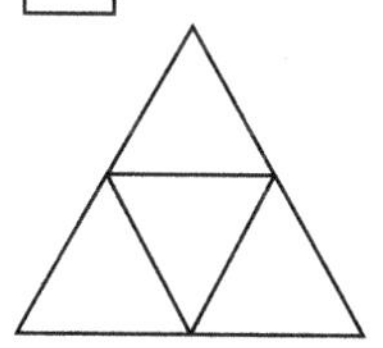

5

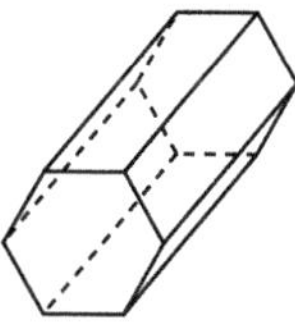

a 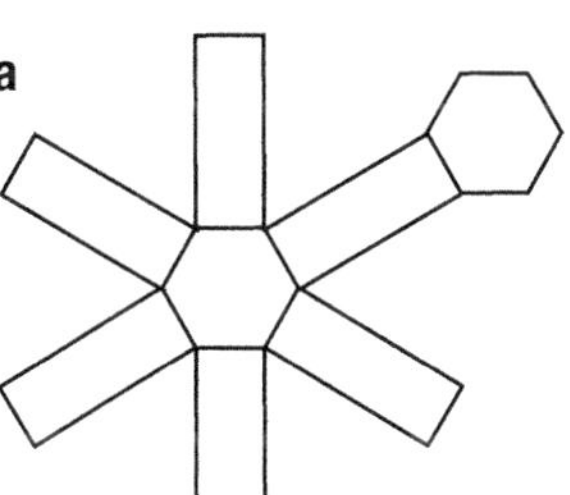

b 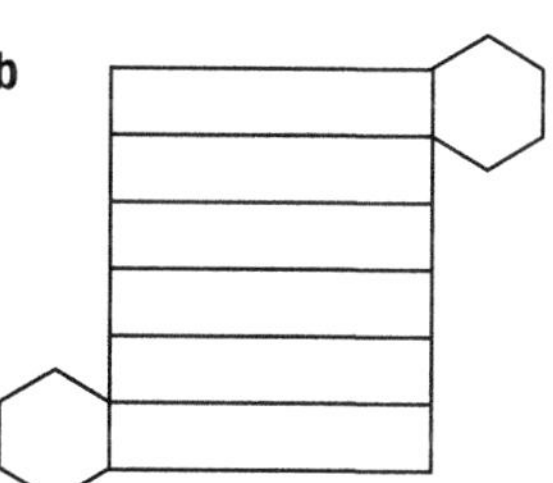

c 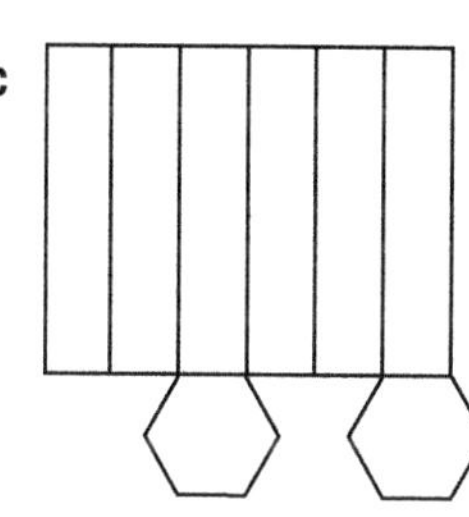

6

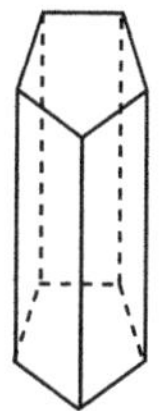

a 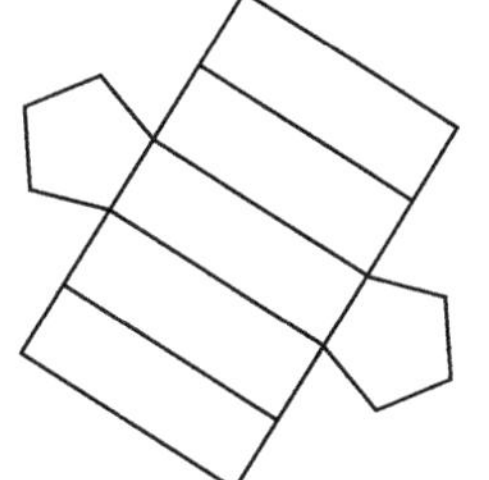

b 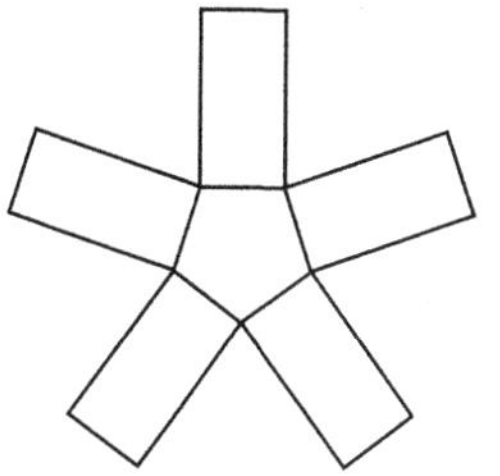

c 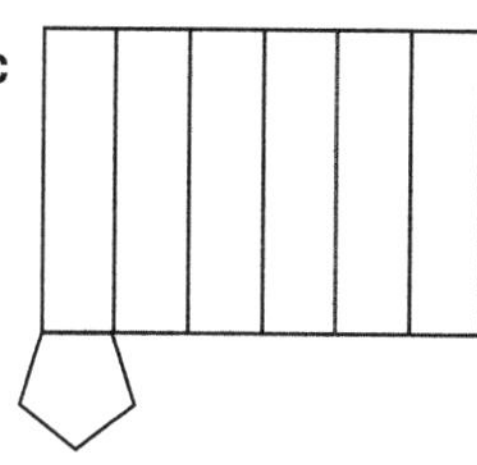

Classify triangles

Which of the following are:

1 equilateral triangles?

2 isosceles triangles?

3 scalene triangles?

4 right-angled triangles?

a

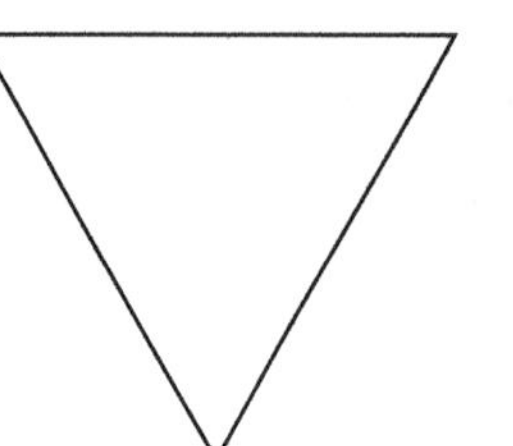

b

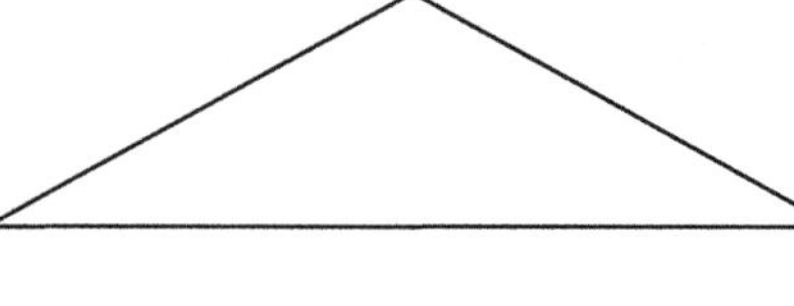

c

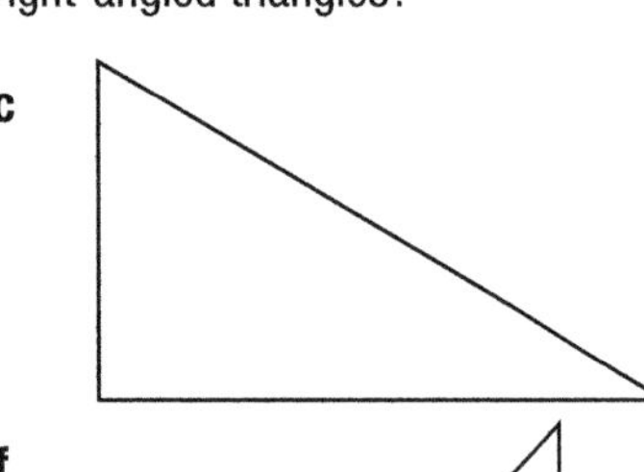

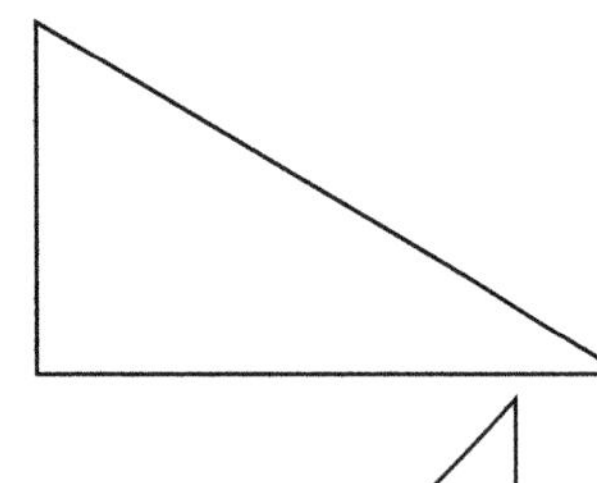

d

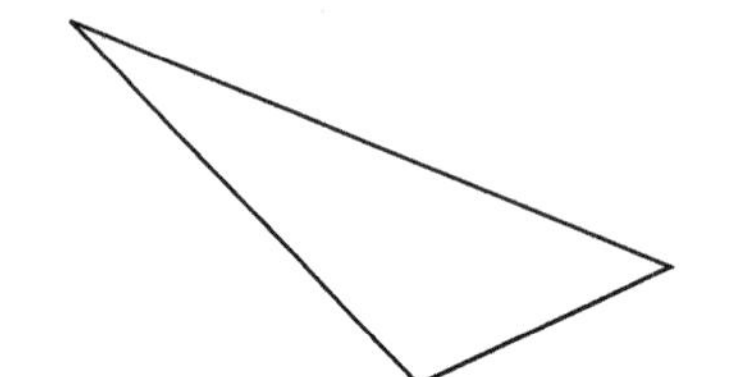

e

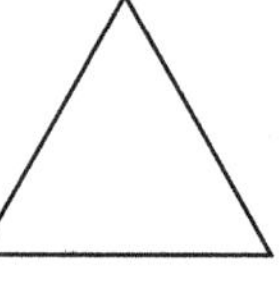

f

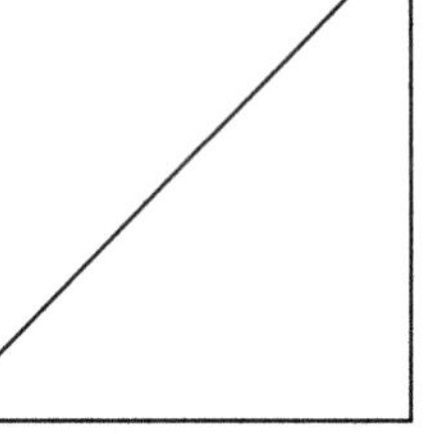

g

j

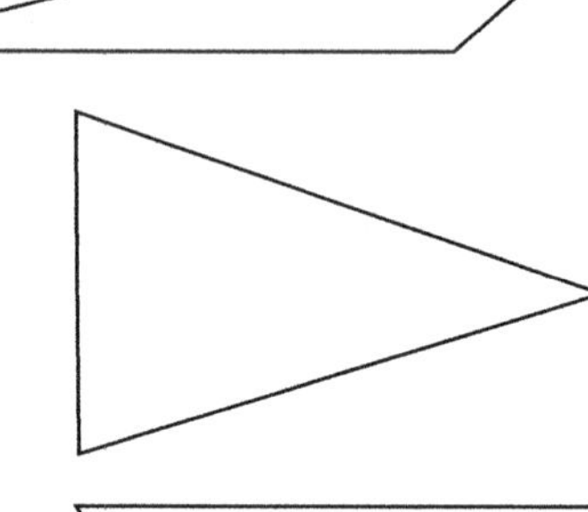

h

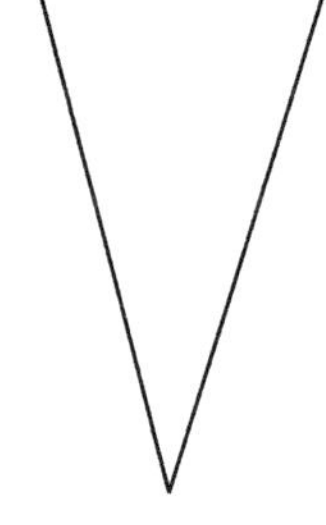

i

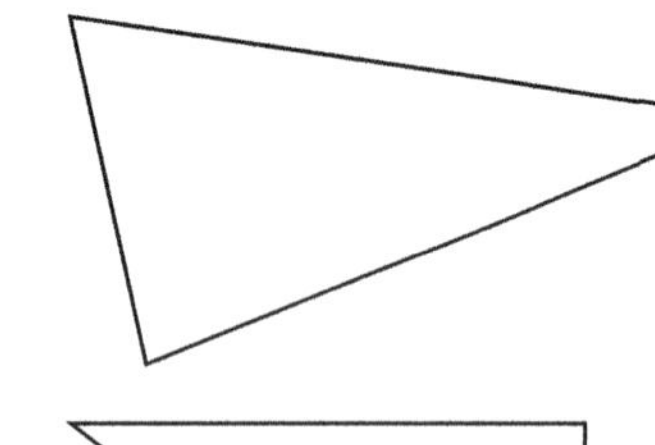

k

m

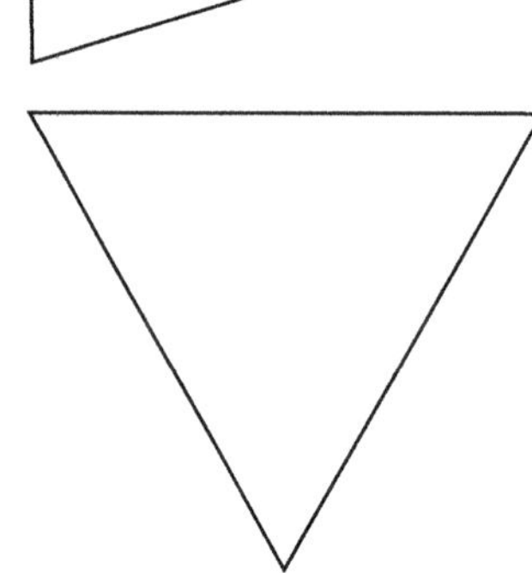

n

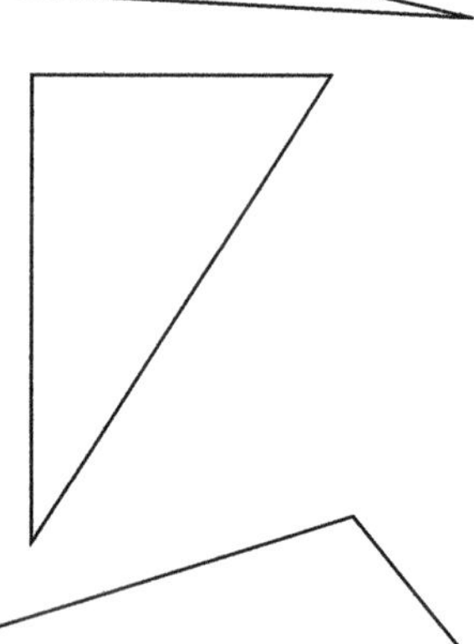

l

o

p

q

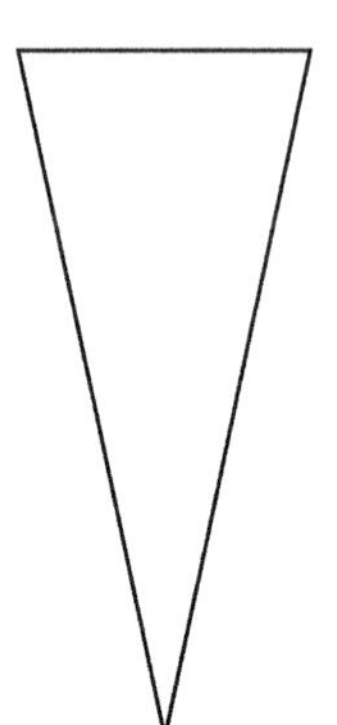

r

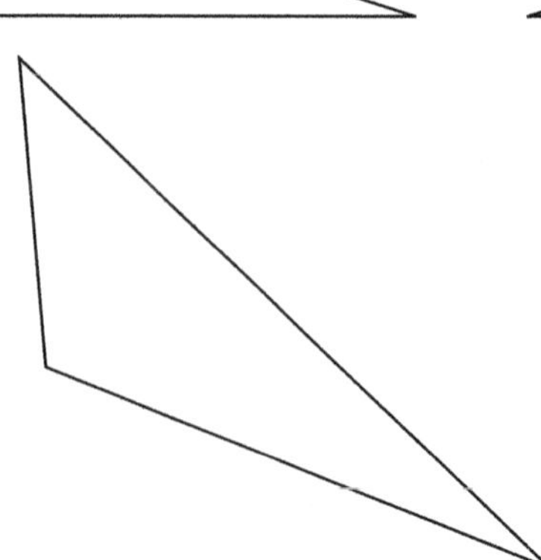

s

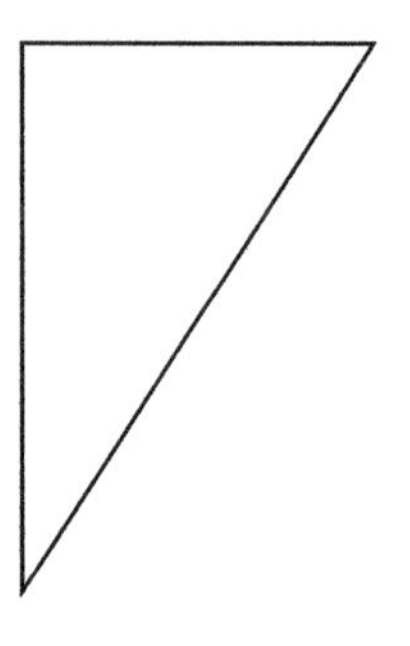

Explore symmetry

Which of the following shapes have:

1. horizontal symmetry?
2. vertical symmetry?
3. horizontal and vertical symmetry?

a

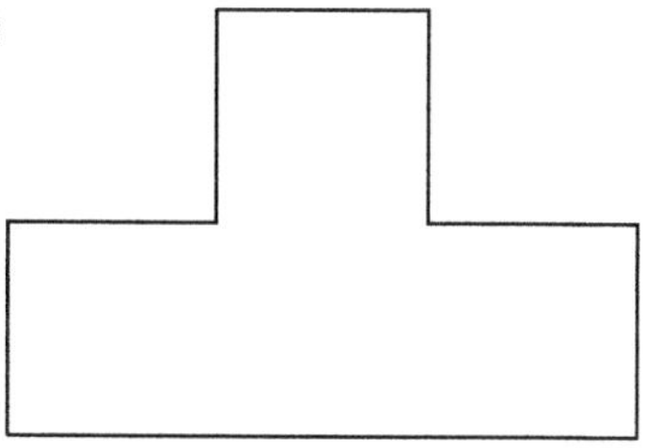

b

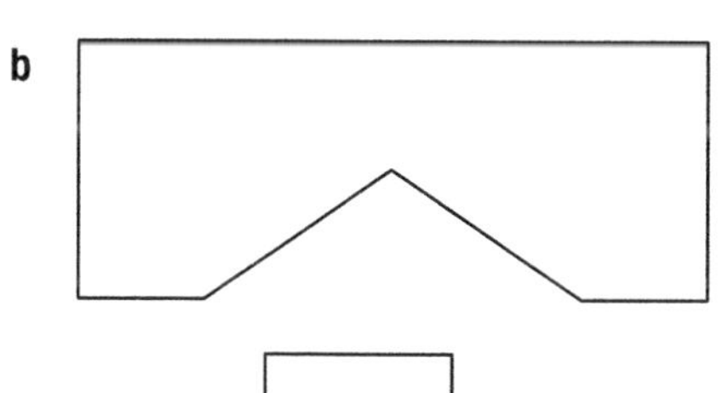

c

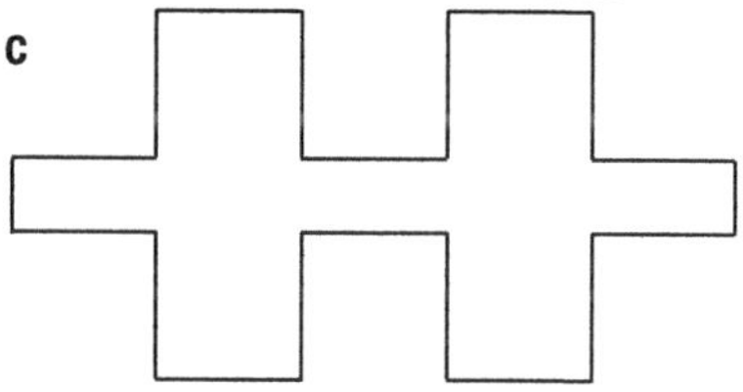

d

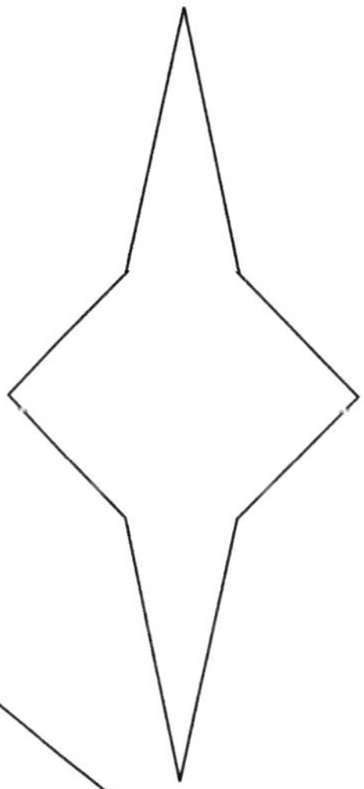

e

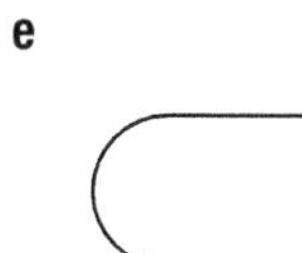

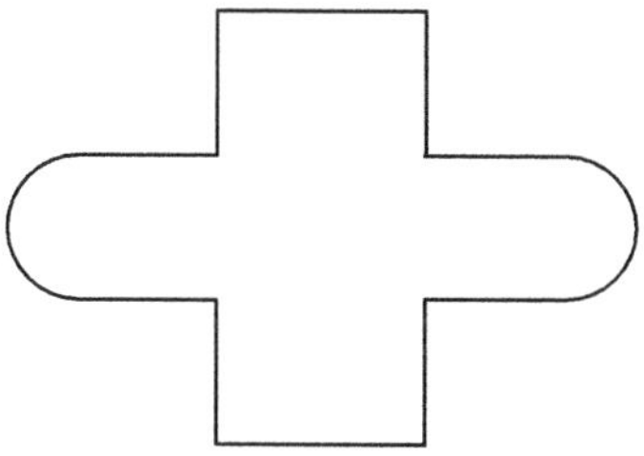

f

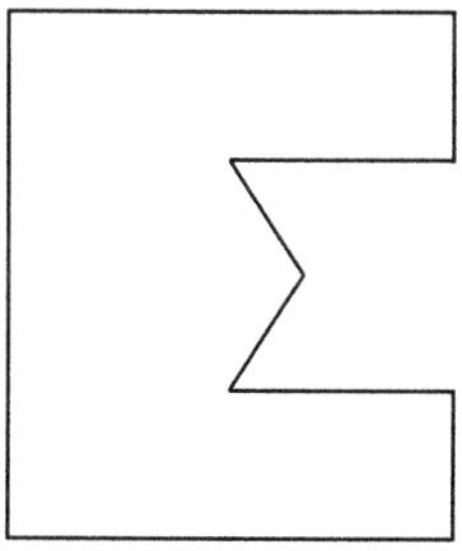

g

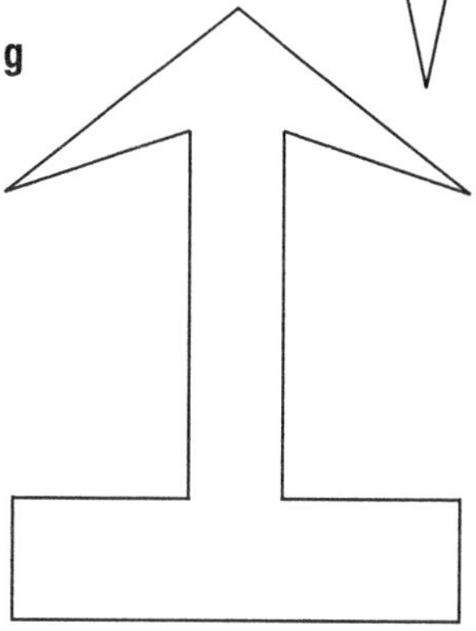

h

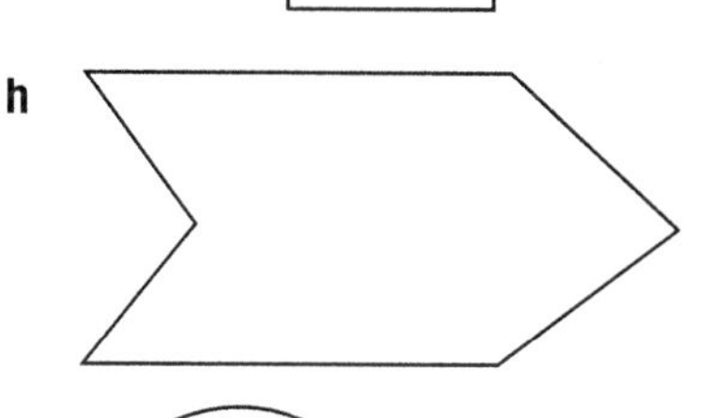

j

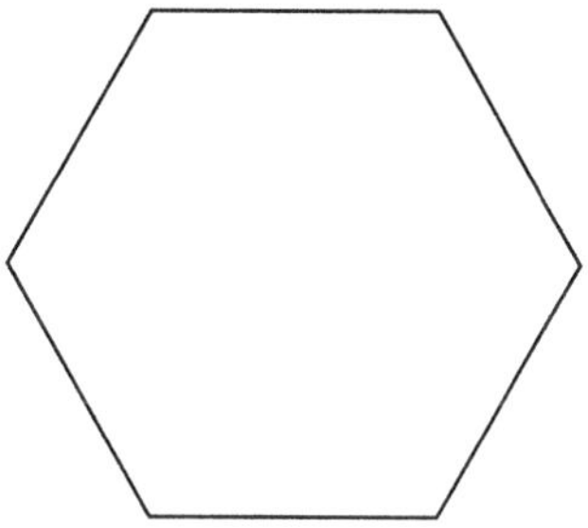

i

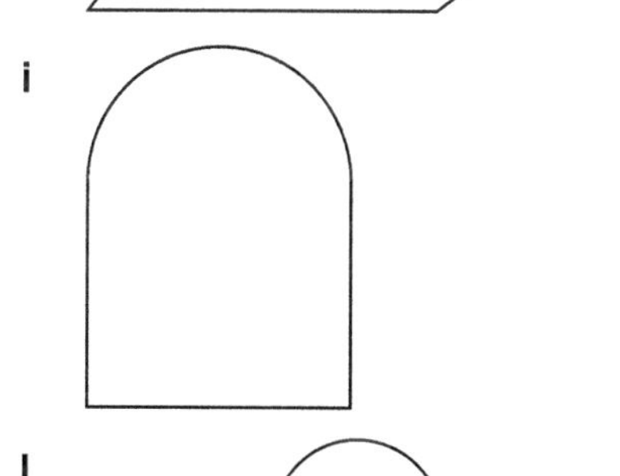

k

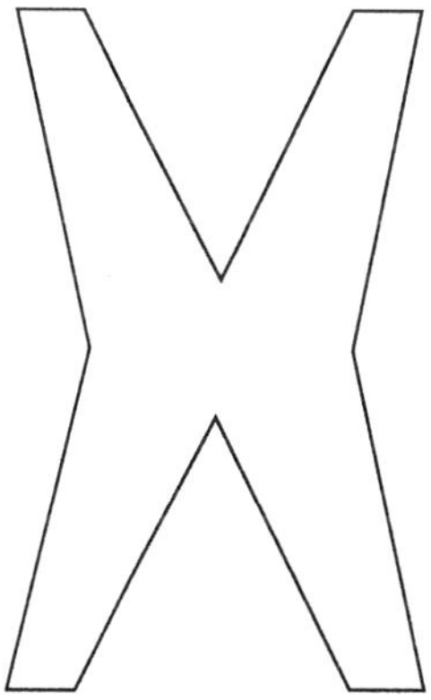

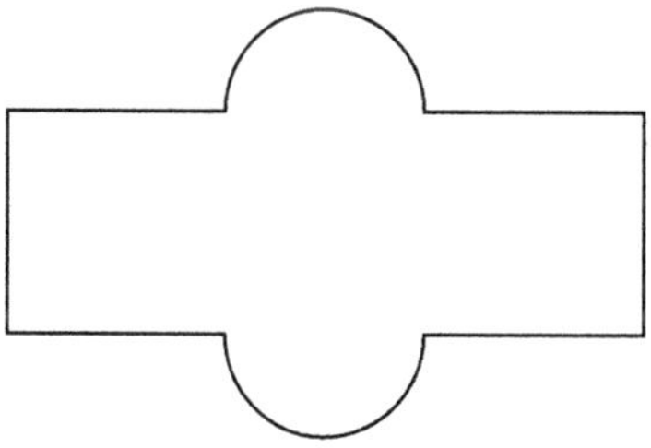

m

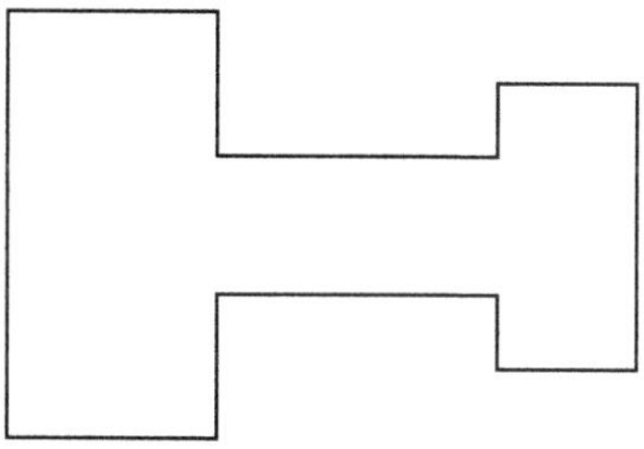

l

o

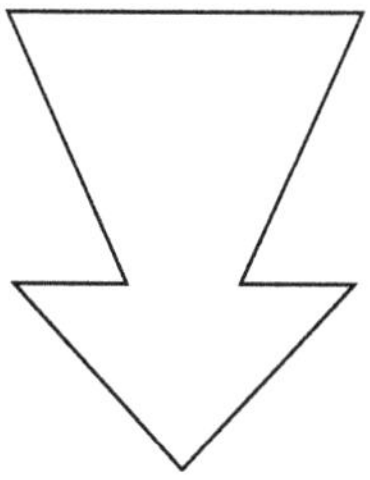

p

n

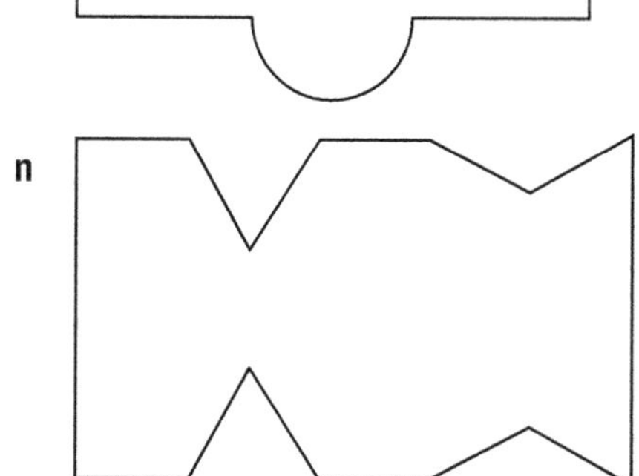

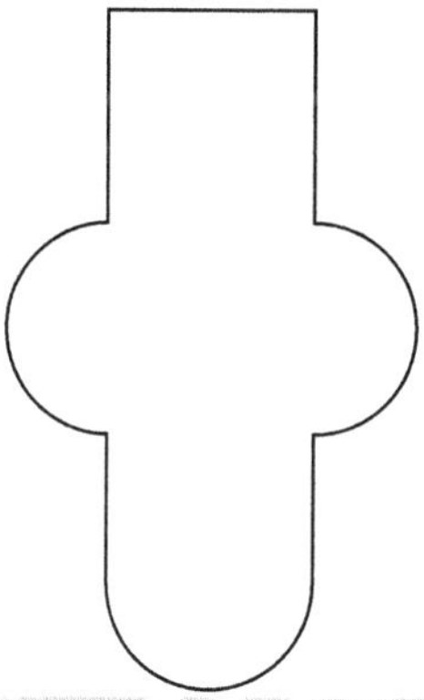

5.3.2 Construct and name angles

Identify and name angles

1 Use acute, obtuse, right, straight or reflex to name each angle.

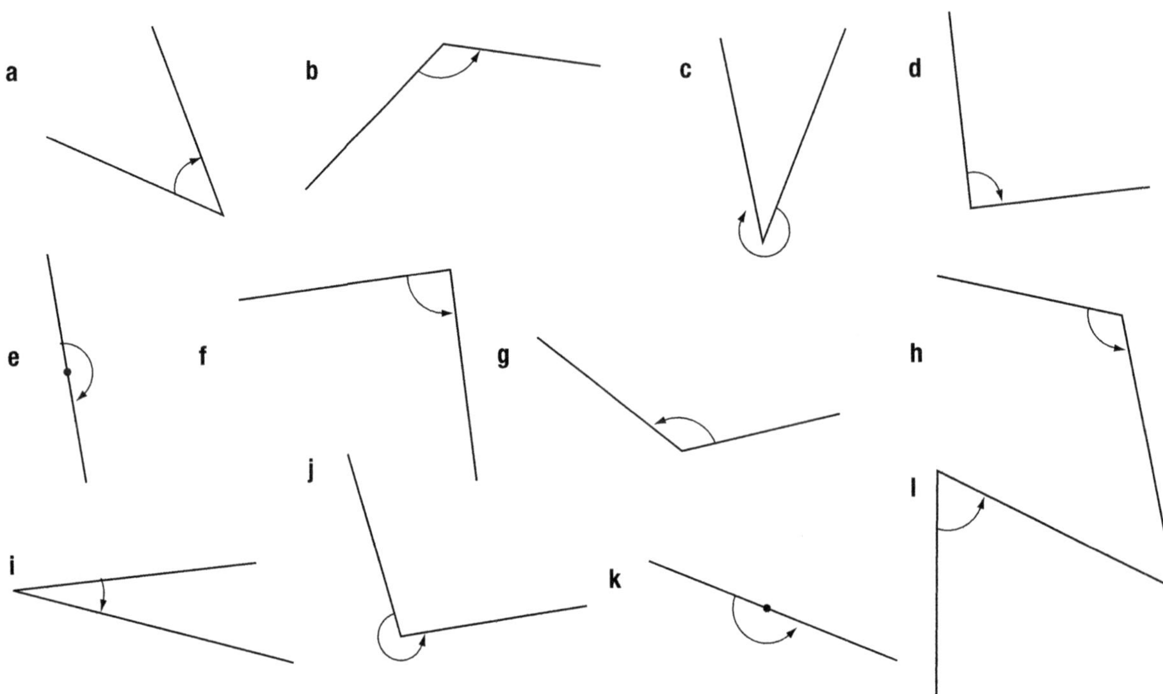

2 Name the angle that is marked in each shape.

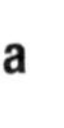

a 

b 

c 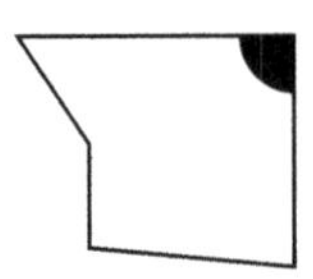

d 

e 

f 

g 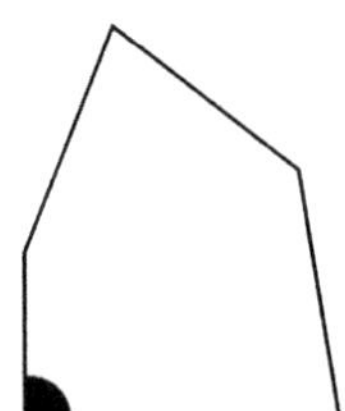

h 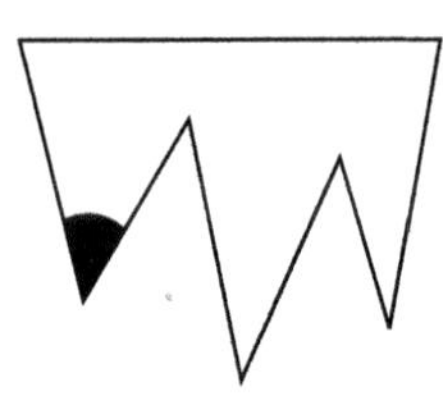

i 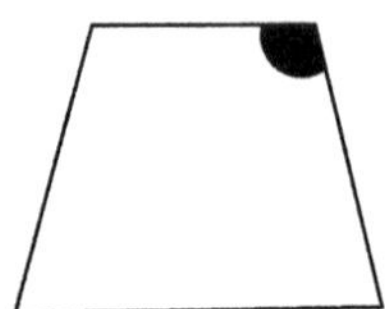

j 

k 

l 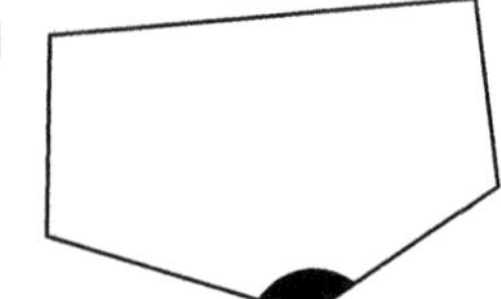

Estimate and measure angle size

Estimate the size of each angle, and then measure each one with a protractor. Record your estimates and measurements in a table like the one shown here with 20 rows.

	Estimate	Measure
1		
2		

Follow compass directions

Copy each grid and then draw the described route, starting at the point marked on each grid.

1

North

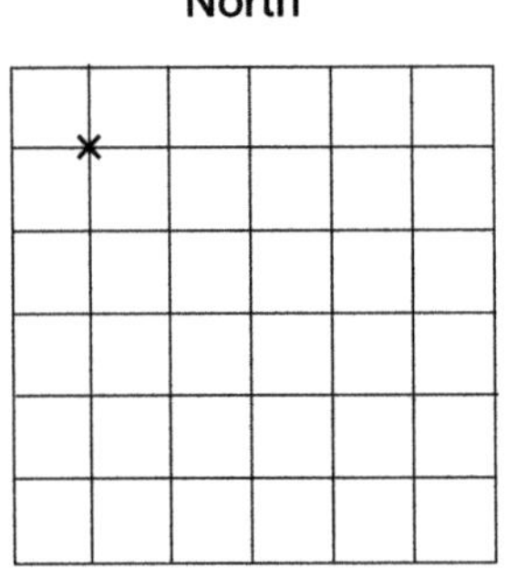

2 south, 3 east, 1 north,
2 west, 3 south, 2 east

2

North

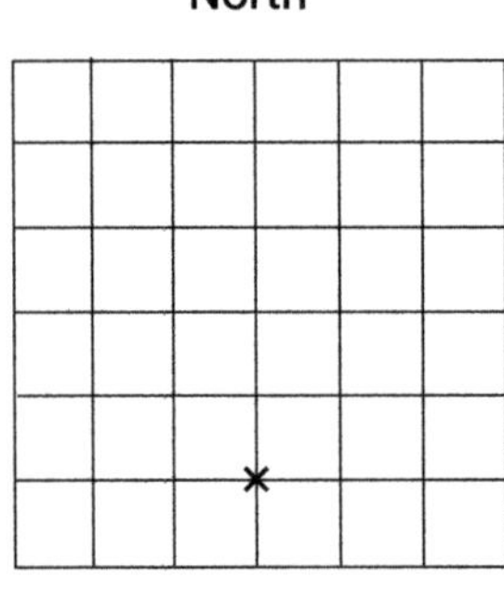

3 north, 1 west, 1 north,
3 east, 3 south, 1 west

3

North

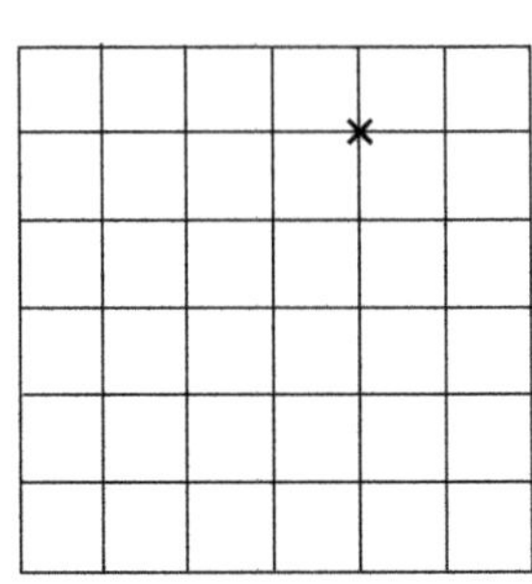

3 west, 4 south, 1 east,
3 north, 3 east, 2 south

4

North

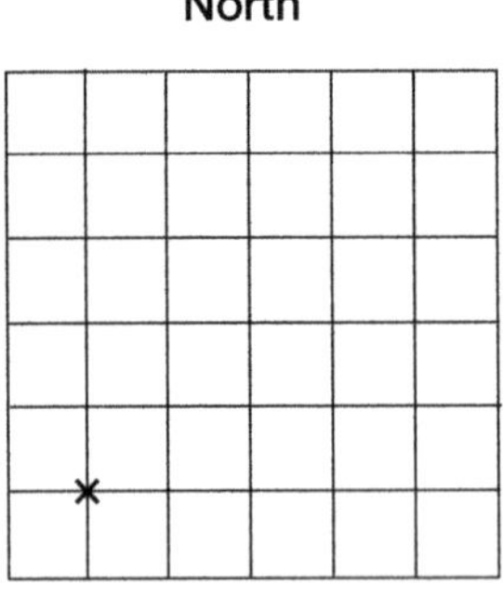

4 east, 3 north, 2 west,
1 north, 3 east, 4 south

5

North

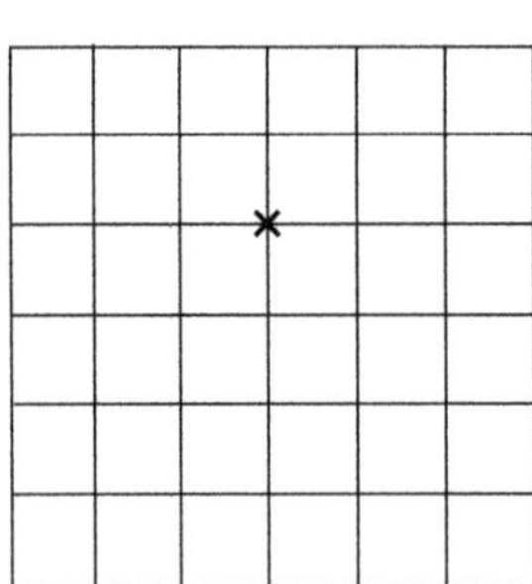

1 north, 3 west, 4 south,
5 east, 2 north, 4 west

6

North

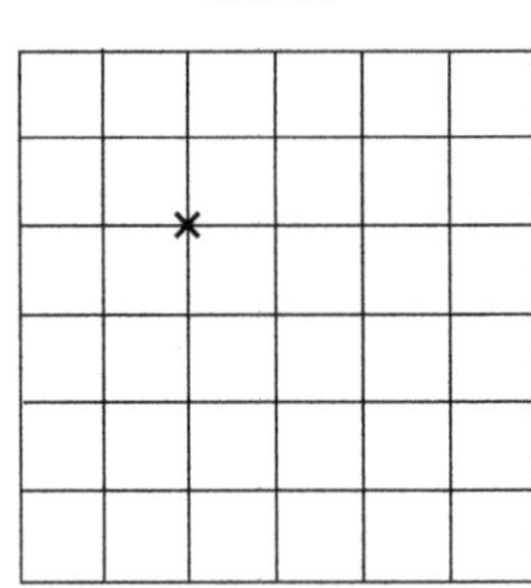

3 south, 3 east, 4 north,
4 west, 3 south, 3 east

7

North

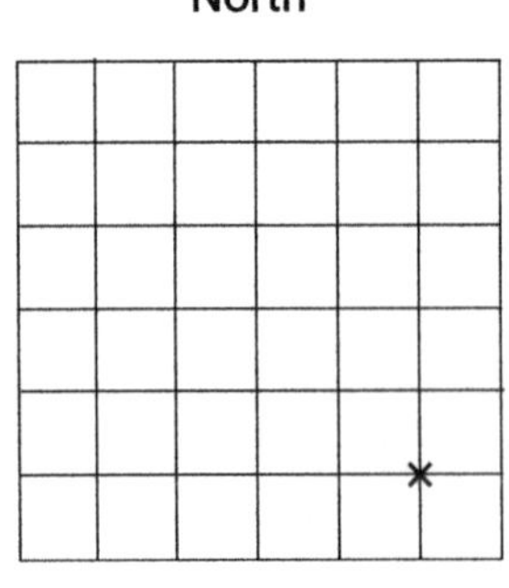

5 west, 2 north, 3 east,
3 north, 2 west, 4 south

8

North

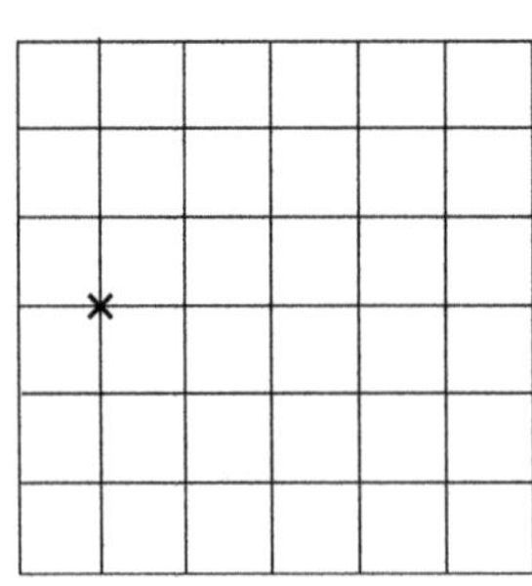

2 south, 4 east, 3 north,
3 west, 2 south, 4 east

9

North

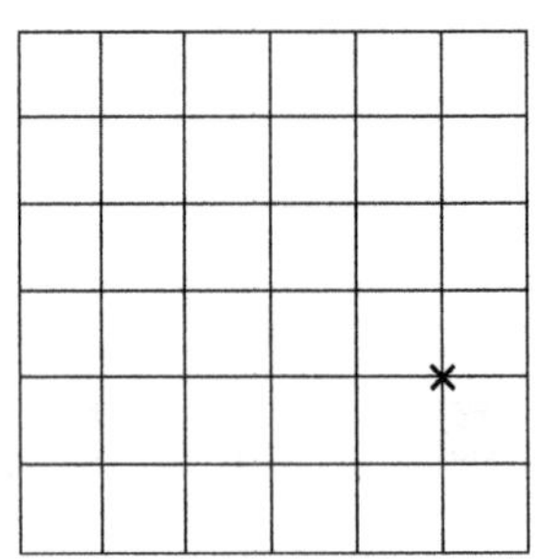

4 north, 4 west, 5 south,
2 east, 4 north, 3 west

Use compass directions to describe routes

Use numbers and compass directions to describe how to get from A to B on each grid. (For example, 2 south, 3 east, 5 north …)

1

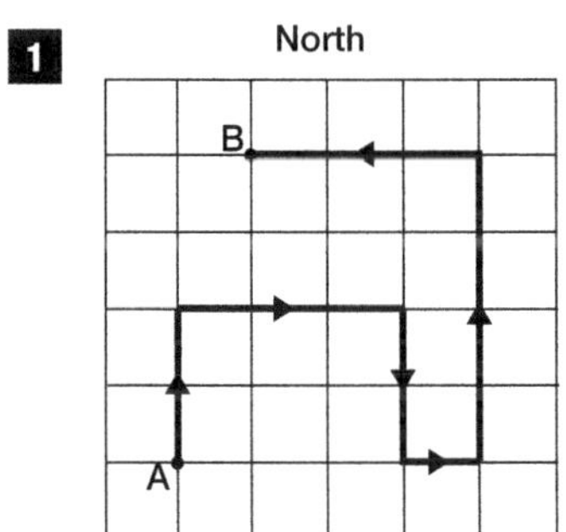

2

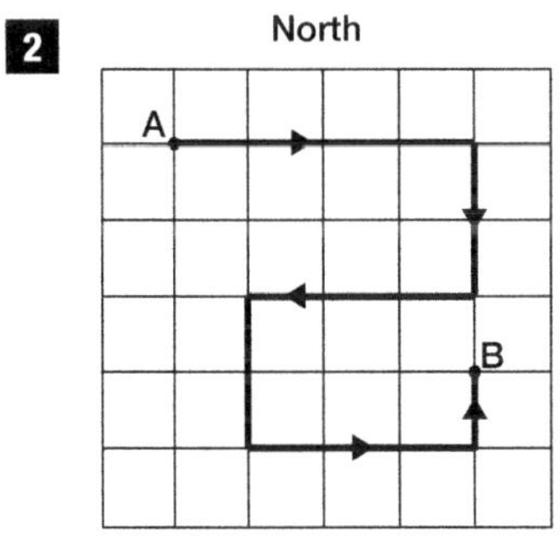

3

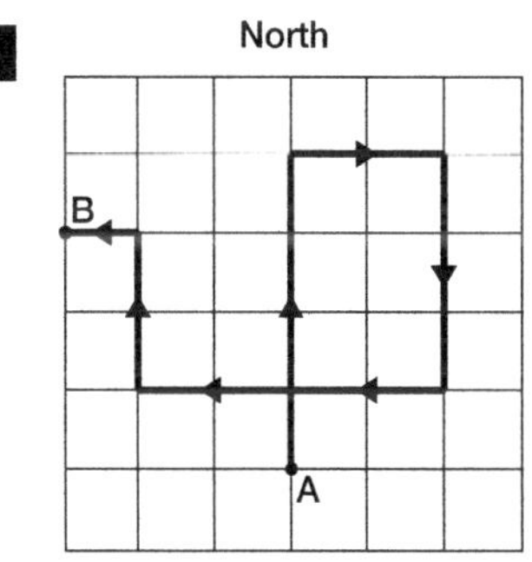

4

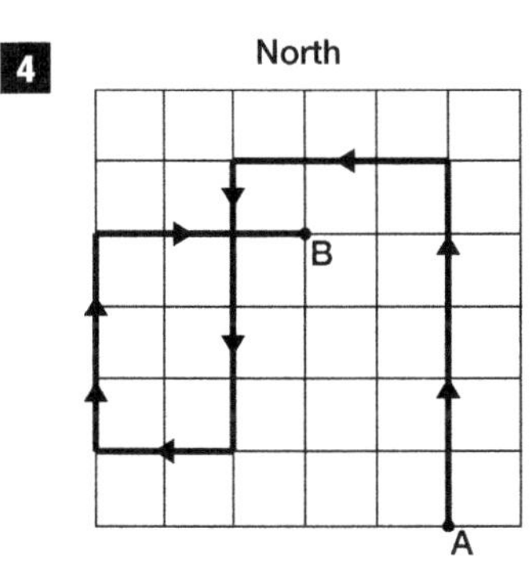

5

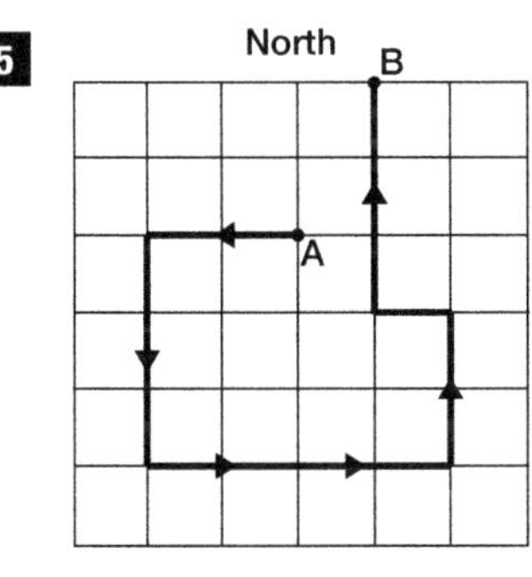

6

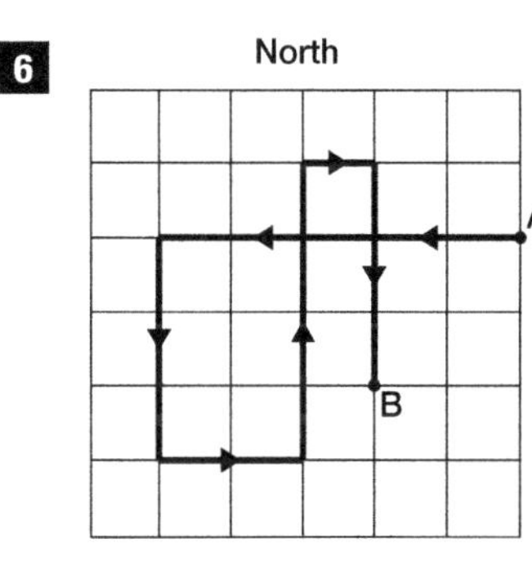

7

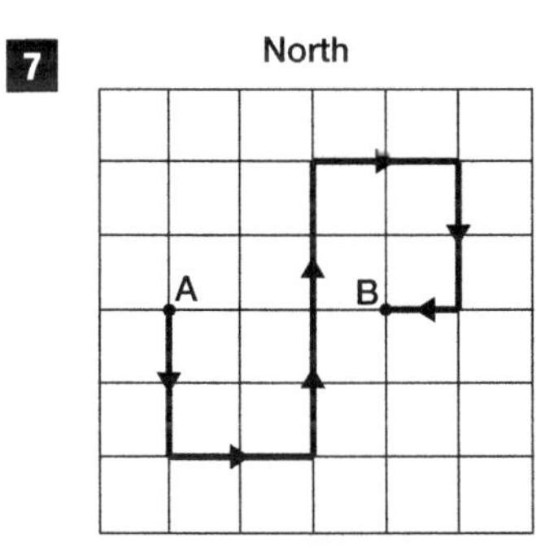

8

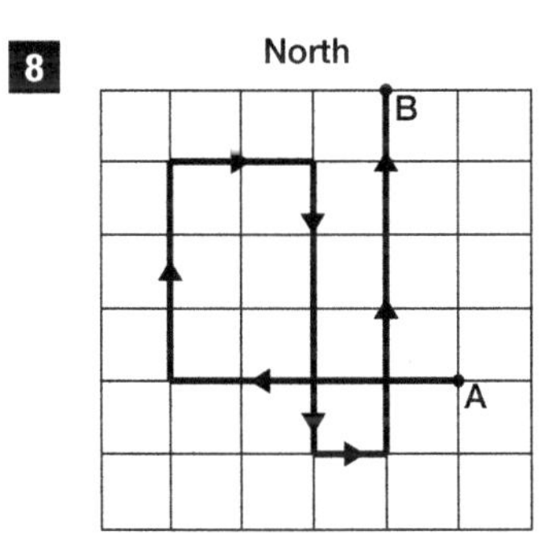

9

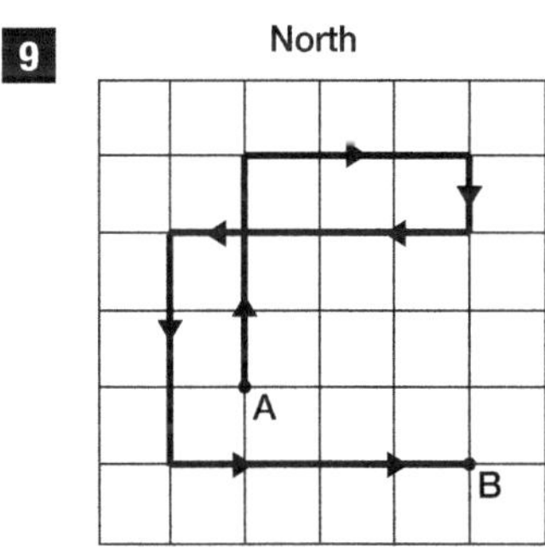

10

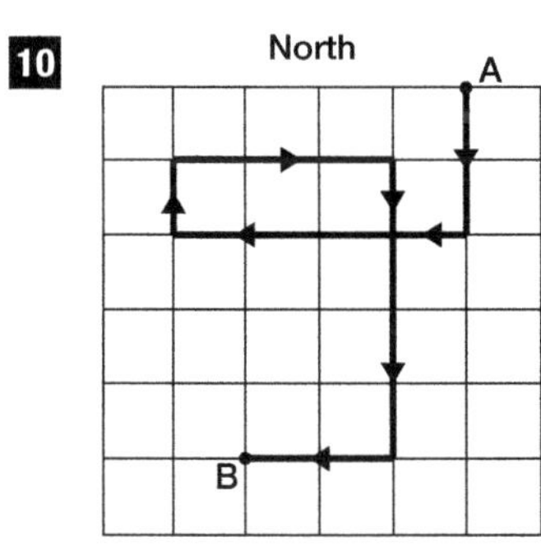

11

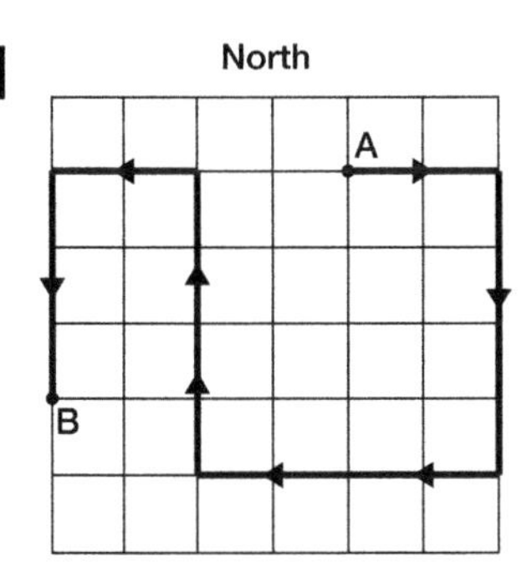

12

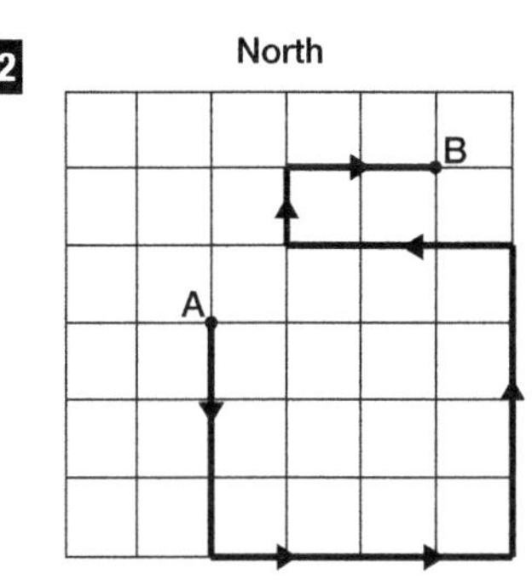

Assessment — Space and Shape

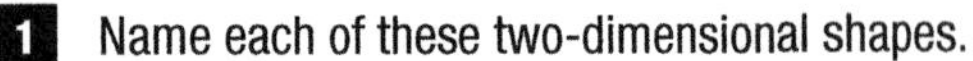

1 Name each of these two-dimensional shapes.

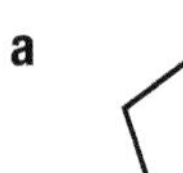

a 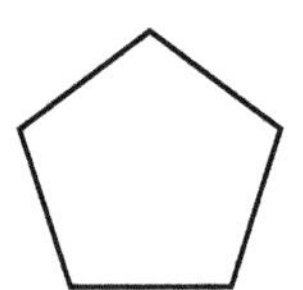 b 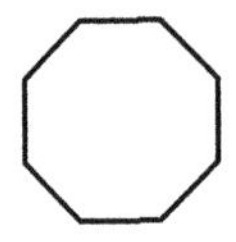 c 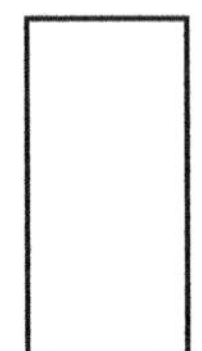 d 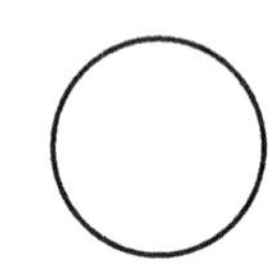 e 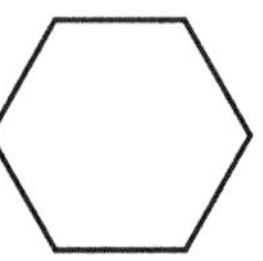

2 Name each of these three-dimensional shapes.

a 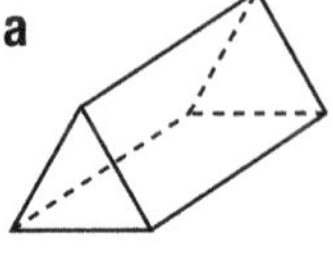 b 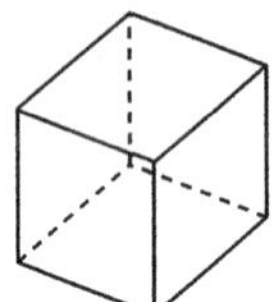 c 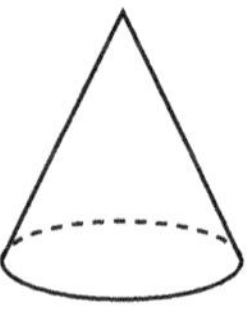 d 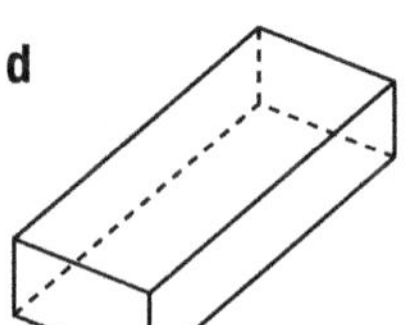 e 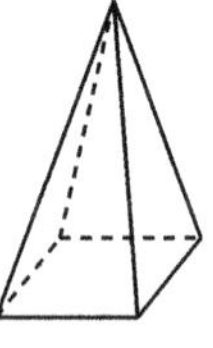

f 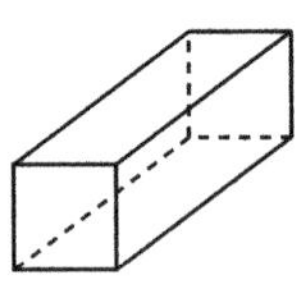 g 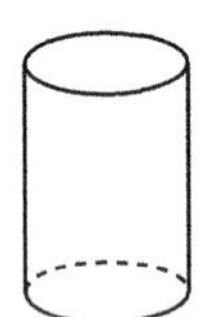 h 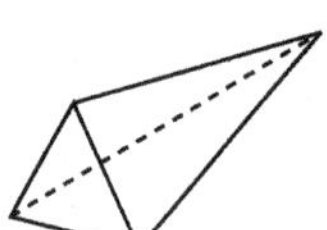 i 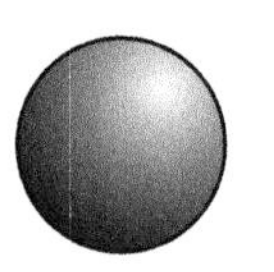 j 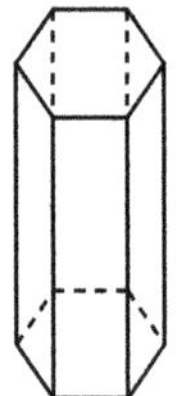

3 Which nets can be made into a triangular prism?

a 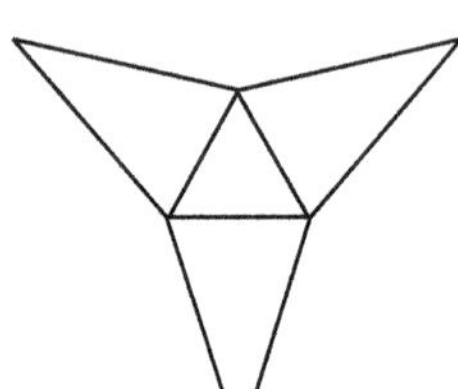 b 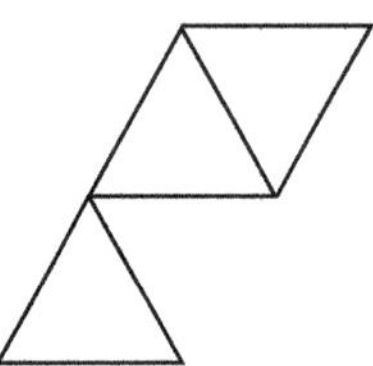 c 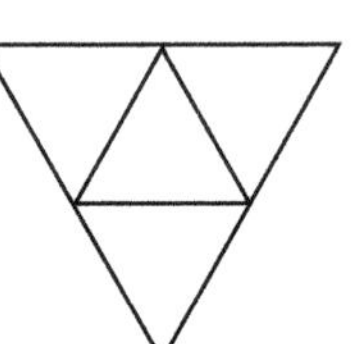 d 

4 Which of the following triangles are equilateral?

a 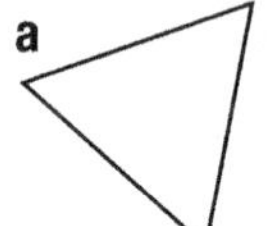 b 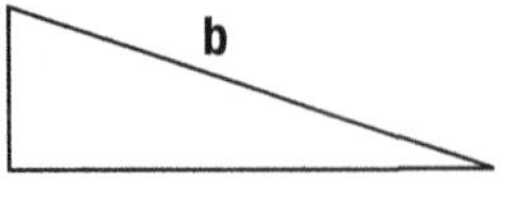 c 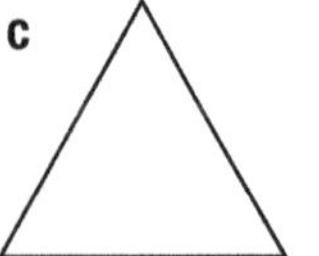 d e 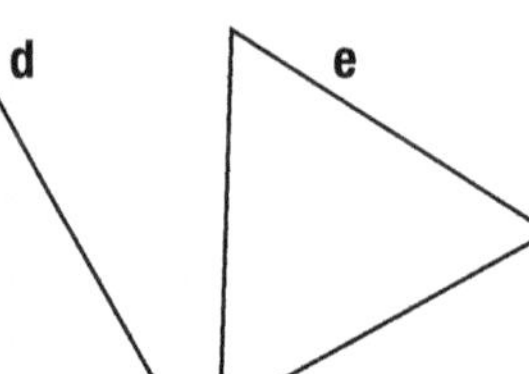 f 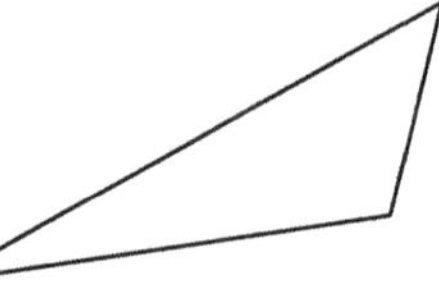

5 Which of these shapes have either horizontal or vertical symmetry?

a  b 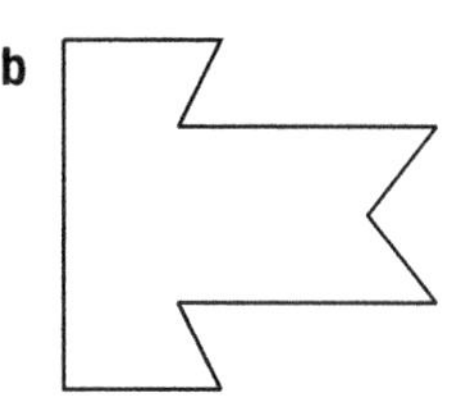 c  d 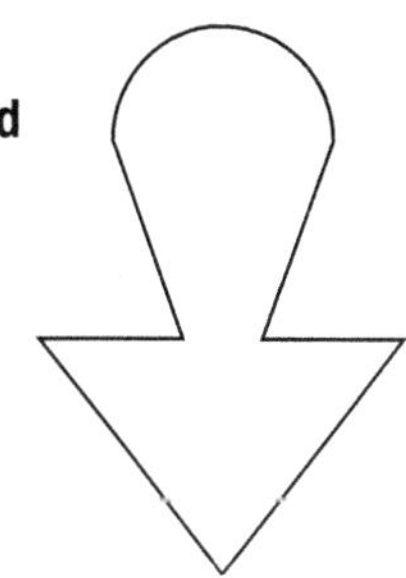 e 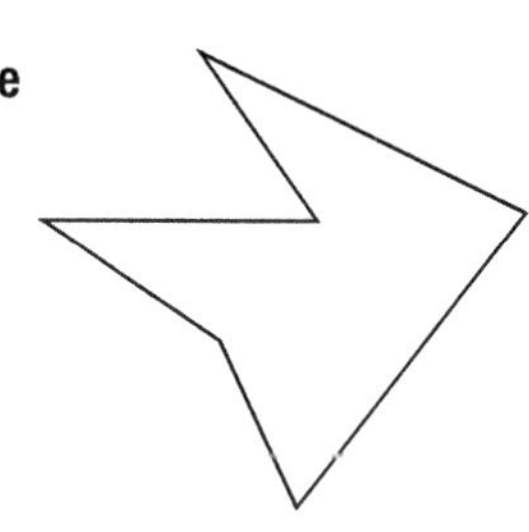

6 Name each angle.

a 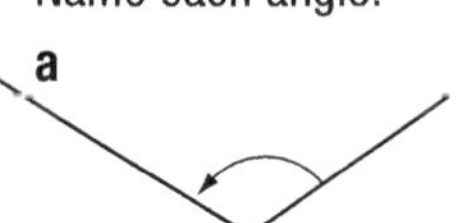 b 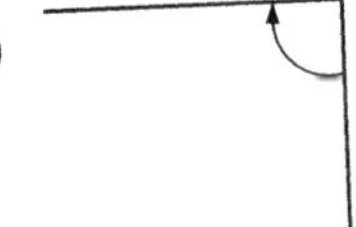 c  d 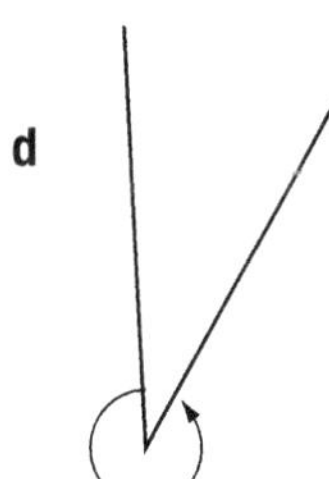 e 

7
a What is the name of this shape?
b How many angles are inside this shape?
c How many of these angles are acute?

8 Use a protractor to measure each angle.

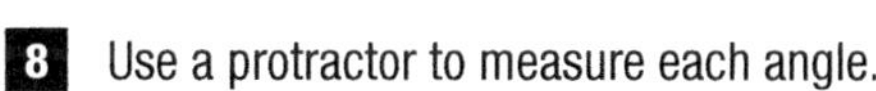

a 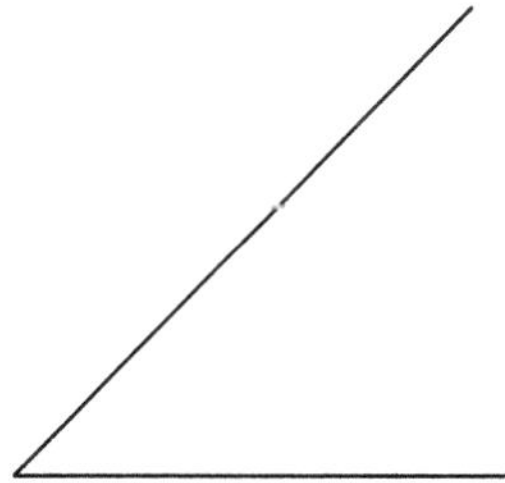  b 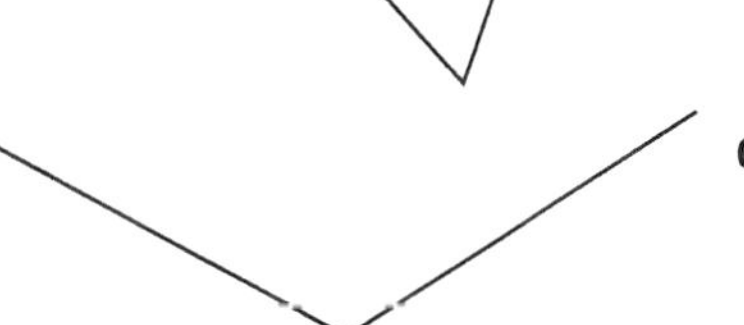 c 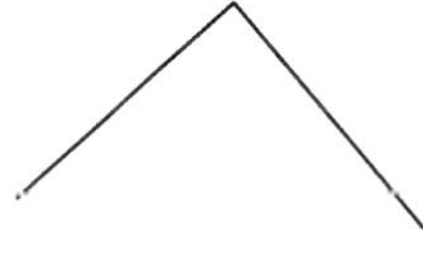

9 Copy the grid and draw the described route starting at the point marked on the grid.

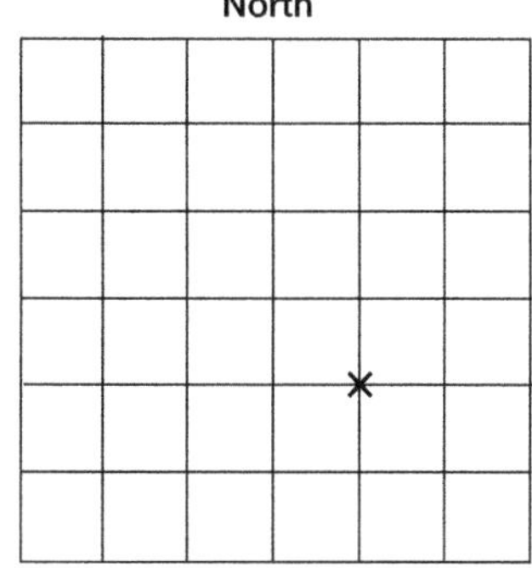

3 north, 3 west, 2 south,
4 east, 2 south, 4 west

10 Use numbers and compass directions to describe how to get from A to B on this grid.

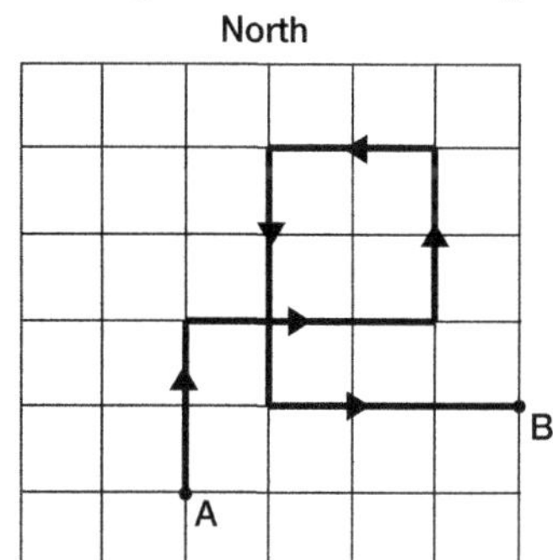

Strand Chance and Data

5.4.2 Represent and interpret information using graphs, tables and charts

Calculate the mean

Help Box

To find the **mean** of a set of scores, first add the scores together and then divide the total by the number of scores.

For example, to find the mean of this set of six scores: 17, 23, 11, 56, 28 and 9
- add the scores together: 144
- divide 144 by the number of scores (6): 144 ÷ 6 = 24

so the mean is 24.

1 Calculate the mean of the following sets of scores.

a 6, 13, 7, 12, 19 and 15

b 12, 25, 21, 17, 34 and 29

c 12, 7, 23, 18, 16, 23 and 20

d 34, 27, 36, 41, 23, 49 and 28

e 26, 17, 5, 31, 46, 21, 36 and 34

f 35, 47, 31, 38, 42, 27, 45 and 23

2 This chart shows the results of eight students for ten tests. Each test was marked out of 20.

Name	Test 1	Test 2	Test 3	Test 4	Test 5	Test 6	Test 7	Test 8	Test 9	Test 10
Elsie	15	18	10	12	19	19	18	20	17	18
Peter	9	13	12	14	14	11	16	12	15	15
Ben	18	18	19	17	17	17	18	20	19	20
Imata	15	15	15	17	16	17	12	15	18	17
Philip	19	20	18	19	20	20	19	19	17	20
Taita	11	16	10	13	15	15	17	15	17	17
Regina	16	16	13	15	17	15	18	18	16	17
Nancy	17	18	18	19	20	20	19	20	18	19

a Find the mean score for each student. Write your answers as decimal numbers where necessary.

b Who has the highest mean score?

c Who has the lowest mean score?

3 This chart shows the highest daily temperature in five places over one week.

Place	Day 1	Day 2	Day 3	Day 4	Day 5	Day 6	Day 7
Port Moresby	30°	30°	31°	34°	31°	32°	34°
Rabaul	31°	32°	30°	31°	31°	31°	30°
Lae	28°	27°	27°	30°	28°	28°	29°
Goroka	26°	26°	25°	27°	24°	26°	25°
Mt. Hagen	21°	24°	22°	21°	21°	22°	24°

a What is the mean weekly temperature for each place? Write your answers as mixed numbers where necessary.

b Which place has the highest mean weekly temperature?

Find the median and mode

1 Find the median of each set of scores.

a 16, 11, 18, 19, 17, 24, 11, 21, 15
b 27, 33, 24, 28, 21, 19, 37, 31, 20
c 56, 52, 59, 44, 49, 51, 45, 55, 50
d 9, 13, 10, 17, 14, 16, 11, 18, 10
e 38, 41, 32, 30, 45, 47, 37, 43, 40
f 64, 57, 61, 62, 52, 59, 63, 51, 68
g 73, 65, 76, 79, 60, 70, 67, 71, 74
h 49, 55, 43, 50, 46, 45, 41, 58, 42
i 34, 29, 37, 24, 28, 21, 32, 30, 27
j 87, 81, 88, 93, 99, 89, 96, 84, 90

Help Box

To find the **median** of a set of scores, first put the scores in order and then find the middle score.

For example: 15, 12, 25, 22 and 19

Put the scores in order: 12, 15, 19, 22, 25

The middle score is 19 so it is the median.

The **mode** of a set of scores is the score that occurs most often.

For example, in the set of scores 6, 8, 4, 8, 5, 8, 4, 7, 9, 8, the number 8 occurs most often, so 8 is the mode.

2 Find the mode of each set of scores.

a 6, 3, 5, 3, 7, 5, 3, 4, 5, 3, 8, 3, 6, 3, 7, 3
b 12, 15, 11, 10, 12, 14, 12, 12, 15, 17
c 19, 14, 17, 18, 16, 18, 19, 18, 14, 18
d 25, 27, 22, 27, 28, 22, 27, 21, 27, 22
e 33, 30, 33, 37, 36, 36, 33, 31, 30, 38
f 85, 87, 80, 81, 78, 88, 81, 76, 81, 72
g 3.6, 2.5, 1.9, 2.5, 2.7, 1.8, 2.5, 1.4, 1.3
h 7.9, 6.1, 7.2, 6.7, 6.1, 7.1, 7.4, 6.1, 7.2
i 5.4, 5.7, 4.8, 5.7, 5.3, 4.5, 4.1, 5.1, 4.2
j 3.7, 3.2, 2.6, 2.2, 2.6, 3.1, 2.9, 3.3, 2.6

3 Find the median of these even sets of scores.

a 7, 4, 8, 9, 11, 6, 15, 5, 10, 3, 12, 9
b 16, 20, 10, 18, 14, 19, 23, 11, 21, 15
c 36, 27, 30, 22, 31, 35, 30, 24, 39, 24
d 23, 34, 38, 21, 25, 35, 20, 29, 37, 33
e 87, 75, 80, 82, 79, 72, 81, 85, 76, 73
f 51, 47, 58, 59, 45, 40, 43, 52, 54, 40
g 63, 67, 60, 69, 75, 70, 77, 71, 64, 76
h 92, 89, 90, 98, 95, 85, 83, 94, 87, 88

Remember

To find the median of an even number of scores, calculate the mean of the middle two scores.

For example, in the scores 9, 12, 13, 15, 18 and 21, the median is the mean of 13 and 15 (28 ÷ 2), so it is 14.

4 Over a six-month period, a store sold pairs of shoes in the following sizes.

7 8 $8\frac{1}{2}$ 10 8 $7\frac{1}{2}$ 6 7 $6\frac{1}{2}$ 9 $7\frac{1}{2}$
8 $7\frac{1}{2}$ 6 9 $5\frac{1}{2}$ 8 5 $7\frac{1}{2}$ 9 $9\frac{1}{2}$ 8
7 $7\frac{1}{2}$ 5 $6\frac{1}{2}$ 8 6 8 $7\frac{1}{2}$ $7\frac{1}{2}$ 9 $6\frac{1}{2}$
10 $8\frac{1}{2}$ 7 $8\frac{1}{2}$ $6\frac{1}{2}$ 7 $7\frac{1}{2}$ 8 8 $7\frac{1}{2}$ $6\frac{1}{2}$ $5\frac{1}{2}$

a Draw a frequency table to show how many pairs of shoes of each size were sold.
b What is the mode of the shoe sizes?
c What size shoes are most popular at this store?

Strand Patterns

5.5.1 Use patterns and arithmetical rules to solve problems

Describe rules

1 Copy and complete each grid by working out the rule that connects the numbers in the top row to the numbers in the bottom row. Describe each rule. (For example, the rule to complete the first grid is 'add 5'.)

a

5	8	3	12	17	9
10	13	8			

b

20	11	18	23	15	19
13	4	11			

c

6	10	7	13	2	25
60	100	70			

d

25	40	15	30	50	100
5	8	3			

e

4	6	3	11	7	20
16	36			49	

f

16	20	8	36	80	28
8	10			40	

g

7	12	25	9	14	31
21	36		27		

h

340	230	510	290	460	370
34	23				37

i

9	15	34	40	28	19
18	30		80		

j

17	14	21	38	26	47
67		71			97

2 Try these harder two-step grids. (For example, the rule to complete the first grid is 'double and add 2'.)

a

4	6	10	5	2	9
10	14	22			

b

3	5	2	10	6	8
10	16		31		

c

8	12	15	9	20	25
		29	17		49

d

9	27	18	30	15	36
4			11	6	

e

16	20	8	24	14	10
7	9			6	

f

60	30	100	70	80	120
		11		9	13

g

5	10	8	20	12	15
13	23	19			

h

30	100	18	60	70	40
17		11	32		

i

60	48	100	26	18	40
27		47		6	

j

6	9	3	7	11	2
65	95		75		

Continue, create and describe number sequences

1 Copy and complete each number sequence.

- **a** 25, 30, 35, 40, ____, ____, ____
- **b** 96, 86, 76, 66, ____, ____, ____
- **c** 20, 21, 23, 26, 30, ____, ____, ____
- **d** 10, 11, 13, 14, 16, ____, ____, ____
- **e** 70, 65, 55, 50, 40, ____, ____, ____
- **f** 1, 3, 9, 27, ____, ____, ____
- **g** 2, 4, 8, 16, ____, ____, ____, ____
- **h** 3, 6, 12, 24, ____, ____, ____, ____
- **i** 32, 30, 29, 27, 26, ____, ____, ____, ____
- **j** 53, 57, 56, 60, 59, ____, ____, ____
- **k** 10, 21, 43, 87, ____, ____, ____
- **l** 3, 5, 9, 17, 33, ____, ____, ____
- **m** 12, 14, 17, 19, 22, ____, ____, ____
- **n** 71, 66, 65, 60, 59, ____, ____, ____
- **o** 1, 2, 6, 12, 36, ____, ____, ____, ____
- **p** 5, 13, 29, 61, ____, ____, ____
- **q** 87, 77, 78, 68, 69, ____, ____, ____, ____
- **r** 9, 19, 18, 28, 27, ____, ____, ____
- **s** 0.2, 0.4, 0.8, 1.6, ____, ____, ____, ____
- **t** 0.3, 0.6, 1.2, 2.4, ____, ____, ____

2 Describe how each number sequence has been created. (For example, 3, 5, 9, 17, 33, 65, 129, 257 has been created by doubling and subtracting 1.)

- **a** 8, 17, 26, 35, 44, 53, 62, 71
- **b** 12, 23, 34, 45, 56, 67, 78, 89
- **c** 2, 8, 16, 64, 128, 512, 1024
- **d** 67, 61, 55, 49, 43, 37, 31, 25
- **e** 45, 43, 40, 38, 35, 33, 30, 28
- **f** 11, 16, 26, 31, 41, 46, 56, 61
- **g** 4, 9, 19, 39, 79, 159, 319, 639
- **h** 7, 13, 25, 49, 97, 193, 385
- **i** 2, 7, 22, 67, 202, 607, 1822
- **j** 83, 73, 68, 58, 53, 43, 38, 28
- **k** 5, 6, 8, 11, 15, 20, 26, 33, 41
- **l** 95, 94, 92, 89, 85, 80, 74, 67
- **m** 3, 12, 11, 20, 19, 28, 27, 36
- **n** 80, 71, 72, 63, 64, 55, 56, 47
- **o** 1.7, 3.4, 6.8, 13.6, 27.2, 54.4
- **p** 1.2, 3.6, 10.8, 32.4, 97.2, 291.6
- **q** 7, 8, 8.5, 9.5, 10, 11, 11.5, 12.5
- **r** 9, 8.8, 8.3, 8.1, 7.6, 7.4, 6.9, 6.7
- **s** 12, 14, 13.5, 15.5, 15, 17, 16.5
- **t** 2.5, 5, 5.5, 11, 11.5, 23, 23.5

3 Create your own number sequences to fit each of these descriptions. You may start at a number of your choice but each sequence must have at least 8 numbers.

- **a** Add 3
- **b** Subtract 7
- **c** Multiply by 2 and then add 1
- **d** Multiply by 2 and then subtract 1
- **e** Add 5 and then add 10
- **f** Subtract 5 and then subtract 10
- **g** Add 7 and then subtract 1
- **h** Subtract 9 and then add 2
- **i** Add 1, then 2, then 3 and so on
- **j** Subtract 2, then 4, then 6 and so on
- **k** Halve
- **l** Multiply by 3 and then subtract 5
- **m** Multiply by 3 and then subtract 10
- **n** Add 8 and then add 3
- **o** Subtract 9 and then subtract 7
- **p** Add 2, then 4, then 6 and so on
- **q** Add 0.5 and then add 1.5
- **r** Add 1.5 and then subtract 0.4
- **s** Subtract 2.5 and then subtract 2
- **t** Multiply by 2 and then subtract 0.5

Solve equations

> **Help Box**
>
> When solving equations, it is usual to do operations in the following order.
> - do calculations in **brackets**
> - work out a fraction **of** a number
> - do **multiplication** and **division** in the order they occur in an equation but before addition and subtraction
> - do **addition** and **subtraction** in the order they occur in an equation and after multiplication and division

1 Use the correct order of operations to solve these equations.

a	$5 \times 8 + 4 - 8$	**b**	$25 - 7 - 3 + 5$		
c	$(3 + 6) \times 5 + 10 - 3$	**d**	$(52 + 8) \times 3 - 7$		
e	$42 \div 6 + 1$	**f**	$(10 + 30) \div 10 - 2$		
g	$25 - (6 + 4) \times 2$	**h**	$3 \times (9 + 9)$		
i	$4 \times 5 + 10$	**j**	$5 \times (4 + 8)$		
k	$23 - 3 \times 3$	**l**	$30 \div 5 + 10$		
m	$20 - 16 \div 4$	**n**	$(18 - 6) \times 3$		
o	$(15 + 8) \times 2$	**p**	$20 + \frac{1}{2}$ of 16		
q	$\frac{1}{2}$ of $10 + 5 \times 3$	**r**	$6 + 3 - (2 + 2)$		
s	$30 - 5 \times 6 + 8$	**t**	$(7 + 3) \times \frac{1}{2}$ of 6	**u**	$18 \div (3 + 3)$
v	$25 - 6 \times 3$	**w**	$(6 - 2) \times (8 - 5)$	**x**	$\frac{1}{4}$ of $12 \times (8 - 6)$

2 Put brackets in these equations to make them correct.

a	$5 \times 6 + 4 = 50$	**b**	$40 - 6 + 4 = 30$	**c**	$10 + 2 \times 5 = 60$
d	$10 + 15 \div 5 = 5$	**e**	$12 - 2 \times 4 = 40$	**f**	$10 - 5 + 3 = 2$
g	$8 \times 4 + 6 = 80$	**h**	$5 \times 8 + 4 - 5 = 55$	**i**	$36 - 6 + 10 = 20$
j	$24 \div 6 + 2 = 3$	**k**	$6 + 8 - 3 + 9 = 2$	**l**	$20 - 4 \times 3 = 48$
m	$\frac{1}{2}$ of $8 + 2 = 5$	**n**	$6 \times 10 - 8 = 12$	**o**	$7 + 3 \times 4 = 40$
p	$5 \times 8 + 2 = 50$	**q**	$20 - 8 \times 4 = 48$	**r**	$30 - 6 \div 2 = 12$
s	$5 \times 3 + 3 - 2 = 28$	**t**	$7 - 3 \times 7 + 2 = 36$	**u**	$20 \div 2 + 2 = 5$
v	$3 + 5 \times 5 - 1 = 32$	**w**	$7 \times 4 + 4 - 6 = 50$	**x**	$18 - 9 \times 10 = 90$

3 Check each equation to see if it is correct or incorrect. Write '=' if it is correct and '≠' if it is incorrect.

a	$15 - 6 \times 2$ ☐ 3	**b**	$4 \times 7 + 5$ ☐ 48	**c**	$10 \div (2 + 3)$ ☐ 2
d	$3 \times (6 + 3)$ ☐ 27	**e**	$5 + 4 + 7 \times 2$ ☐ 32	**f**	$25 \div 5 - 2$ ☐ 3
g	$16 + 3 \times 8$ ☐ 40	**h**	$(5 + 6) \times 2$ ☐ 17	**i**	$(15 - 5) \times 8$ ☐ 80
j	$(3 + 6) \times (10 - 5)$ ☐ 45	**k**	$7 + 8 \times 2$ ☐ 30	**l**	$21 - 5 \times 2$ ☐ 32
m	$\frac{1}{2}$ of $10 + 8$ ☐ 9	**n**	$50 - 24 \div 4$ ☐ 44	**o**	$16 + \frac{1}{4}$ of 8 ☐ 18
p	$4 + 6 \times 7$ ☐ 46	**q**	$12 - 5 \times 2$ ☐ 14	**r**	$3 \times 4 + 5 \times 2$ ☐ 22
s	$6 + 4 \times 4$ ☐ 40	**t**	$(8 - 2) \times 8$ ☐ 48	**u**	$7 \times 3 - 10$ ☐ 11
v	$(16 + 4) \div 5$ ☐ 4	**w**	$3 \times 3 + 6 - 2$ ☐ 21	**x**	$32 \div 4 - 2 \times 2$ ☐ 4

4 Solve these more difficult equations.

a	$\frac{1}{4}$ of $20 + 8 - \frac{1}{2}$ of 6	**b**	$50 - 48 \div 8 + 15$	**c**	$7 + 8 \times 5 - \frac{1}{4}$ of 12
d	$(17 + 12) \times 2 + 10 - 5$	**e**	$3 \times (20 - 8) + \frac{1}{2}$ of 16	**f**	$56 \div 7 + 45 \div 9 - \frac{1}{4}$ of 8

Assessment — Chance and Data and Patterns

Multiple choice test for Graphs and tables (5.4.2) and Patterns (5.5.1)

1 What is the mean of 17, 14, 20, 11 and 8?

a 20 b 14 c 17 d 70

2 What is the mean of 22, 15, 25, 31, 7, 24 and 37?

a 21 b 20 c 23 d 22

3 What is the mean of 19, 16, 26, 9, 21 and 17?

a 17 b 18 c 19 d 20

4 What is the median of 6, 9, 3, 15 and 7?

a 7 b 6 c 9 d 8

5 What is the median of 53, 71, 60, 58, 67, 63, 75, 55 and 70?

a 53 b 75 c 60 d 63

6 What is the median of 43, 52, 39, 49, 55 and 34?

a 55 b 49 c 46 d 43

7 What is the median of 3, 11, 7, 18, 14, 9, 13 and 8?

a 9 b 10 c 11 d 12

8 What is the mode of 5, 7, 9, 3, 6, 7, 8, 7, 9, 3, 2, 7 and 5?

a 3 b 5 c 7 d 9

9 What is the mode of 12, 17, 16, 12, 14, 15, 12, 18, 17 and 11?

a 12 b 13 c 14 d 15

10 What is the mode of $7\frac{1}{2}$, 8, 8, $7\frac{1}{2}$, $8\frac{1}{2}$, 8, $7\frac{1}{2}$, $8\frac{1}{2}$ and $7\frac{1}{2}$?

a $7\frac{1}{2}$ b 8 c $8\frac{1}{2}$ d 7

11 If the rule that connects pairs of numbers is 'subtract 7', which pair of numbers is incorrect?

a 12, 5 b 10, 3 c 15, 8 d 20, 15

12 If the rule that connects pairs of numbers is 'multiply by 3', which pair of numbers is incorrect?

a 4, 12 b 5, 15 c 2, 9 d 7, 21

13 What is the next number in this sequence: 10, 20, 40, 80, 160, …?

a 180 b 320 c 300 d 200

14 What is the next number in this sequence: 4, 6, 7, 9, 10, 12, 13, …?

a 14 b 16 c 15 d 12

15 What is the next number in this sequence: 76, 71, 61, 56, 46, 41, …?

a 40 b 31 c 36 d 51

16 The answer to the equation $10 + 3 \times 6 - 2$ is:

a 76 b 52 c 20 d 26

17 The answer to the equation $30 - (6 + 8) \div 2$ is:

a 23 b 8 c 16 d 28

18 The answer to the equation $\frac{1}{2}$ of $18 + 9 - \frac{1}{4}$ of 12 is:

a 12 b $10\frac{1}{2}$ c 15 d 14

19 The answer to the equation $8 + 15 + 10 \times 2$ is:

a 66 b 56 c 58 d 43

20 The answer to the equation $9 + 2 \times \frac{1}{4}$ of 16 is:

a 17 b 44 c 15 d 25

Important Facts

Number

Place value

The value of a digit in a number is determined by its place value or position.

Tens of thousands	Thousands	Hundreds	Tens	Ones	Tenths	Hundredths	Thousandths

Prime numbers

Numbers that have only two factors: the number itself and 1. The number 1 is not a prime number. The number 2 is the only even prime number. The prime numbers between 0 and 100 are: 2, 3, 5, 7, 11, 13, 17, 19, 23, 29, 31, 37, 41, 43, 47, 53, 59, 61, 67, 71, 73, 79, 83, 89 and 97.

Composite numbers

Numbers that have more than two factors. For example, 4 has 1, 2 and 4 as factors; 12 has 1, 2, 3, 4, 6 and 12 as factors.

Cubed numbers

Cubed numbers are the result of numbers that have been multiplied three times.

For example: $8 = 2 \times 2 \times 2$; $27 = 3 \times 3 \times 3$; $64 = 4 \times 4 \times 4$.

Triangular numbers

All triangular numbers can be arranged in a triangular pattern of dots. For example:

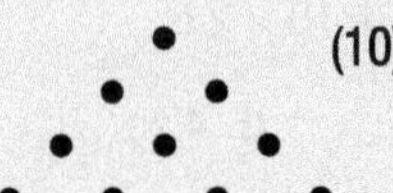

(10)

Triangular numbers from 1 to 100 are 1, 3, 6, 10, 15, 21, 28, 36, 45, 55, 66, 78 and 91.

Square numbers

All square numbers can be arranged in a square pattern of dots.

For example:

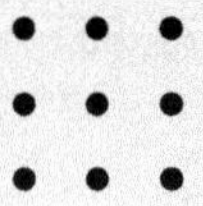

(9 = 3 × 3)

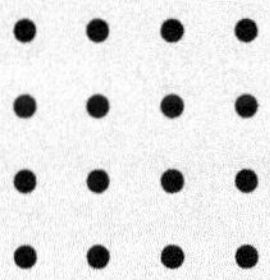

(16 = 4 × 4)

(25 = 5 × 5)

Order of operations

Do **brackets** first, then **of**, then **multiplication** and **division** in the order they appear from left to right, and finally **addition** and **subtraction** in the order they appear from left to right.

Measurement

Length and perimeter

Millimetres, centimetres, metres and kilometres are used to measure length.

10 millimetres (mm) = 1 centimetre (cm)

100 centimetres (cm) = 1 metre (m)

1000 metres (m) = 1 kilometre (km)

1 mm = $\frac{1}{10}$ cm = 0.1 cm

1 cm = $\frac{1}{100}$ m = 0.01 m

1 m = $\frac{1}{1000}$ km = 0.001 km

Time

60 seconds = 1 minute

60 minutes = 1 hour

24 hours = 1 day

7 days = 1 week

2 weeks = 1 fortnight

52 weeks = 1 year

12 months = 1 year

365 days = 1 year

366 days = 1 leap year

10 years = 1 decade

100 years = 1 century

a.m. = ante meridiem – the period between 12 midnight and 12 noon

p.m. = post meridiem – the period between 12 noon and 12 midnight

12-hour time divides a day into two 12-hour periods: a.m. and p.m.

24-hour time divides a day into twenty-four 1-hour periods, starting at midnight

Capacity

Millilitres and litres are used to measure capacity.

1000 millilitres (mL) = 1 litre (L)

1 mL = $\frac{1}{1000}$ L = 0.001 L

Weight

Grams, kilograms and tonnes are used to measure weight.

1000 grams (g) = 1 kilogram (kg)

1000 kilograms (kg) = 1 tonne (t)

1 g = $\frac{1}{1000}$ kg = 0.001 kg

1 kg = $\frac{1}{1000}$ t = 0.001 t

Money

100 toea (t) = 1 kina (K)

1t = $\frac{1}{100}$ K = K0.01

Area

Square centimetres and square metres are used to measure area.

10 000 square centimetres (cm^2) = 1 square metre (m^2)

Area of a rectangle = Length × Width

Volume

Cubic centimetres and cubic metres are used to measure volume.

100 000 cubic centimetres (cm^3) = 1 cubic metre (m^3)

Volume of a rectangular prism = Length × Width × Height

Answers

Strand: Number and Application

Page 2

1 a 7436 b 2758 c 5971 d 9320
e 3897 f 15682 g 46065 h 24107
i 70549 j 68293 k 37350 l 59005

2 a six thousand, seven hundred and eighty-three
b three thousand and eighty-one
c nine thousand, seven hundred and seventy-five
d two thousand, one hundred and fifty
e five thousand, four hundred and forty-two
f one thousand, two hundred and nine
g sixteen thousand, five hundred and fifty-two
h eighty-one thousand, three hundred and forty
i thirty-three thousand, seven hundred and six
j fifty thousand and twenty-seven
k twenty-eight thousand, six hundred and forty-three
l sixty-two thousand, one hundred and fourteen

3 a nine thousand, four hundred and six
b two thousand, five hundred and fifty
c eight thousand, nine hundred and sixty-three
d five thousand, eight hundred and eighteen
e three thousand, four hundred and ninety-seven
f ten thousand, one hundred and four
g forty-three thousand, three hundred and seventeen
h twelve thousand and fifty-six
i ninety thousand, seven hundred and forty-five
j twenty-six thousand, six hundred and eight
k twenty thousand, five hundred and sixty
l seventy-one thousand and thirty-one

4 a 6894 b 2067 c 11943 d 35175 e 59309 f 42217

5 a five thousand, six hundred and seventy kina
b three thousand and ninety-six kina
c fifteen thousand, nine hundred and eighty-five kina
d sixty-three thousand, one hundred and forty-seven kina
e forty-nine thousand, six hundred and seventy-eight kina
f fifty-four thousand and thirty-two kina
g nineteen thousand, two hundred and ninety-nine kina
h thirty thousand and eighty kina
i eighty-five thousand, two hundred kina
j twenty-seven thousand, two hundred and forty-eight kina
k seventy-three thousand, one hundred and fifty-five kina
l ninety-one thousand, two hundred and eighteen kina

6 a K25290 b K70465 c K18725 d K58043

Pages 3–4

1 9890, 11265, 9950, 10063, 9880, 19218, 9910, 9900, 10125

2 15000, 10175, 12758, 15110, 15010, 15115, 14999, 15099, 10512

3 K97115, K79100, K81000, K97182, K96765, K89506, K78060

4 K19995, K20100, K35510, K34800, K35549, K35050, K33550, K30500

5 11755, 9599, 12080, 10999, 9810, 9625, 12010, 9758, 11006

6 20600, 23079, 19785, 22162, 21738, 19792, 23016

7 a – l Teacher to check.

8 a 4556, 5465, 5645, 6445, 6554
b 3018, 3080, 3081, 3180, 3810
c 8034, 8043, 8340, 8403, 8430
d 7129, 7219, 7912, 9127, 9721
e 12037, 12370, 12373, 12703, 12705
f 31060, 31680, 31688, 31806, 31866
g 23129, 23192, 23219, 23291, 23912
h 59007, 59010, 59017, 59070, 59710

9 a – h Teacher to check.

10 a 85500 b 98000 c 55758 d 82000 e 80003 f 37750
g 50696 h 92506 i 62715 j 68796

11 a > b < c > d > e > f <
g < h > i > j > k < l >
m > n < o < p <

Pages 5–7

1 a 2678 b 4582 c 9321 d 1706 e 17948
f 81253 g 62570 h 36407 i 23985 j 57516

2 a 2547 b 8173 c 24316 d 69258 e 91702
f 85345 g 73064 h 62149 i 11980 j 28593

3 a 4500 b 4512 c 45076 d 49666
e 49828 f 40339

4 a 1209 b 8769 c 33229 d 15093
e 57192 f 17639

5 a – l Teacher to check.

6 a 6000 b 12405 c 21007

7 a 3850 b 50063 c 15108

8 a 100 b 10 c 100 d 1000 e 1000
f 10 g 10 h 100 i 1000 j 100
k 10 l 1000 m 10 n 100 o 100

9 a 96275 b 67423, 65725 c 96275, 48265
d 14826, 38620, 67423, 65725, 70629 e 18002, 25612
f 52067, 48265 g 38620, 25612, 70629 h 52067

10 a 5818 b 9206 c 6530 d 28105 e 94429 f 32233
g 44058 h 77999 i 17136 j 59725 k 87040 l 54817
m 64406 n 23325 o 19671 p 50136 q 76578 r 49006
s 60000 t 85063 u 93747 v 44155 w 26510 x 26004

11 a 86431 b 13468 c 31468 d 61348

12 a 97521 b 12579 c 71259 d 51279

Pages 8–9

1 a K90 b K60 c K80 d K40 e K50 f K50
g K130 h K710 i K320 j K560 k K110 l K290
m K410 n K530 o K870 p K160 q K290 r K380
s K2550 t K1340 u K5680 v K3080 w K1900 x K4110

2 a K200 b K200 c K600 d K400 e K100 f K300
g K900 h K400 i K600 j K800 k K500 l K700
m K4400 n K1800 o K2700 p K5100 q K1200 r K3000
s K5900 t K3900 u K7200 v K4700 w K1100 x K2600

3 a 4000 b 6000 c 8000 d 2000 e 2000 f 6000
g 58 000 h 81 000 i 39 000 j 62 000 k 49 000 l 11 000
m 26 000 n 80 000 o 35 000 p 17 000 q 28 000 r 54 000
s 73 000 t 60 000 u 43 000 v 15 000 w 68 000 x 32 000

4 a 90 000 b 50 000 c 70 000 d 30 000 e 80 000 f 20 000
g 10 000 h 80 000 i 50 000 j 20 000 k 90 000 l 40 000
m 40 000 n 10 000 o 30 000 p 70 000 q 70 000 r 30 000
s 90 000 t 50 000 u 60 000 v 10 000 w 90 000 x 70 000

5 6582, 7019, 7499, 6985, 7312, 6500

6 27 500, 30 675, 25 500, 31 275, 32 742, 26 841, 29 036

7 a 14 680, 14 700, 15 000, 10 000
b 53 090, 53 100, 53 000, 50 000
c 38 710, 38 700, 39 000, 40 000
d 22 540, 22 500, 23 000, 20 000
e 85 180, 85 200, 85 000, 90 000
f 68 830, 68 800, 69 000, 70 000
g 12 420, 12 400, 12 000, 10 000
h 75 310, 75 300, 75 000, 80 000
i 47 690, 47 700, 48 000, 50 000
j 91 760, 91 800, 92 000, 90 000

Page 10

1 Prime = 37, 17, 31
Triangular = 45, 36, 10
Odd = 37, 45, 17, 25, 31
Square = 64, 36, 25
Composite = 64, 45, 36, 10, 25, 26
Even = 64, 36, 10, 26

2 64, 9, 25, 16, 100, 81, 36

3 27, 125, 1000, 64, 343

4 5607, 3021, 7555, 8743, 3117, 3159, 14 325, 56 713, 51 617, 88 995

5 **b, c, e, f, h, l, k, m, n, p** will give answers that are even

6 **b, e, f, g, h, i, j, l** will give answers that are odd

Pages 11–12

1 7579, 7587, 7595, 7603

2 4072, 4065, 4058, 4051

3 9703, 9707, 9711, 9715

4 5898, 5907, 5916, 5925

5 6101, 6095, 6089, 6083

6 8019, 8027, 8035, 8043

7 3889, 3894, 3899, 3904

8 5913, (5920), 5927, 5934, 5941

9 3007, (3016), 3025, 3034, 3043

10 7825, (7831), 7837, 7843, 7849

11 4976, (4981), 4986, 4991, 4996

12 6214, 6223, 6232, 6241

13 12 154, 12 149, 12 144, 12 139

14 21 113, 25 122, 25 131, 25 140

15 9992, 9984, 9976, 9968

16 36 524, 36 528, 36 532, 36 536

17 47 145, 47 137, 47 129, 47 121

18 23 115, (23 125), 23 135, 23 145, 23 155

19 39 079, 39 085, 39 091, 39 097

20 56 007, (55 998), 55 989, 55 980, 55 971

21 19 459, (19 466), 19 473, 19 480, 19 487

22 75 627, 75 634, 75 641, 75 648

23 41 867, 41 861, 41 855, 41 849

24 68 238, (68 243), 68 248, 68 253, 68 258

Page 13

1 a 32 b 34 c 35 d 40 e 40
f 40 g 46 h 50 i 40 j 45
k 31 l 30

2 a 25 b 36 c 35 d 40 e 48
f 48 g 46 h 41 i 32 j 47
k 66 l 61

3 a

+	85	78	64	103	96
9	94	87	73	112	105
7	92	85	71	110	103
11	96	89	75	114	107
8	93	86	72	111	104

b

+	126	98	75	110	87
20	146	118	95	130	107
15	141	113	90	125	102
19	145	117	94	129	106
25	151	123	100	135	112

c

+	105	93	138	112	146
14	119	107	152	126	160
18	123	111	156	130	164
23	128	116	161	135	169
40	145	133	178	152	186

d

+	99	152	106	122	163
16	115	168	122	138	179
21	120	173	127	143	184
33	132	185	139	155	196
24	123	176	130	146	187

e

+	101	97	149	58	210
19	120	116	168	77	229
49	150	146	198	107	259
29	130	126	178	87	239
33	134	130	182	91	243

f

+	197	205	161	299	119
11	208	216	172	310	130
9	206	214	170	308	128
51	248	256	212	350	170
99	296	304	260	398	218

Page 14

1 a 395 b 399 c 382 d 376 e 384 f 379

2 a 167 b 553 c 283 d 687 e 619 f 229

Page 15

1 a

×	6	3	7	9	4	8
5	30	15	35	45	20	40
7	42	21	49	63	28	56
4	24	12	28	36	16	32
8	48	24	56	72	32	64
2	12	6	14	18	8	16
6	36	18	42	54	24	48

b

×	5	2	9	11	7	12
3	15	6	27	33	21	36
9	45	18	81	99	63	108
7	35	14	63	77	49	84
6	30	12	54	66	42	72
8	40	16	72	88	56	96
10	50	20	90	110	70	120

2 a

÷	24	36	12	48	18	30	42
3	8	12	4	16	6	10	14
6	4	6	2	8	3	5	7

b

÷	32	48	16	40	72	56	24
4	8	12	4	10	18	14	6
8	4	6	2	5	9	7	3

3 a 5, 25, 5, 30, 6, 36, 4 b 8, 24, 12, 60, 10, 40, 8
c 4, 16, 2, 12, 48, 6, 18 d 2, 24, 4, 36, 6, 48, 4
e 6, 18, 9, 54, 6, 12, 36 f 10, 30, 10, 20, 4, 32, 8
g 8, 16, 4, 28, 7, 49, 7 h 3, 24, 6, 12, 36, 4, 72

Page 16

1 a 160 b 160 c 80 d 2 e 1500 f 1200
g 6 h 80 i 210 j 900 k 7 l 70
m 7 n 2000 o 270 p 3

2 a 6 b 7200 c 50 d 4900 e 4800 f 7
g 1200 h 80 i 9 j 3600 k 560 l 9
m 70 n 2400 o 2500 p 10

3 a 130 b 30 c 120 d 40 e 20 f 160
g 110 h 40 i 120 j 30 k 80 l 170
m 20 n 80 o 110 p 140

4 a 110 b 160 c 30 d 150 e 70 f 90
g 120 h 110 i 80 j 90 k 100 l 110
m 30 n 40 o 20 p 180

5

×	7	4	8	6	9	11	5	10	3	12
30	210	120	240	180	270	330	150	300	90	360
50	350	200	400	300	450	550	250	500	150	600
90	630	360	720	540	810	990	450	900	270	1080
40	280	160	320	240	360	440	200	400	120	480
70	490	280	560	420	630	770	350	700	210	840
80	560	320	640	480	720	880	400	800	240	960

Page 17

Game

Page 18

1 a 1247 b 823 c 1371 d 811 e 1292 f 1116

2 a 1921 b 2377 c 1947 d 2717 e 1776 f 1450

3 a 109 b 290 c 413 d 89 e 1250 f 698
g 1100 h 9500 i 239 j 575 k 525 l 399
m 4400 n 599 o 787 p 2400 q 314 r 563
s 1000 t 796 u 545 v 599 w 416 x 477

4 a 100 b 75 c 900 d 897 e 10 f 40
g 7100 h 400 i 5 j 3006 k 1300 l 9
m 40 n 720 o 995 p 100 q 20 r 354
s 300 t 14 u 15 v 10 w 1992 x 10 000

5 a less than 1000 b more than 1000 c more than 1000
d more than 1000 e more than 1000 f less than 1000
g less than 1000 h less than 1000 i less than 1000
j more than 1000 k more than 1000 l more than 1000

Page 19

1 a 21 b 61 c 45 d 52 e 11 f 200
g 17 h 32 i 32 j 39 k 64 l 5
m 21 n 56 o 300 p 72

2 a 12 b 150 c 80 d 49 e 29 f 38
g 35 h 39 i 37 j 8 k 67 l 450
m 90 n 9 o 64 p 25

3 a 47 b 85 c 68 d 125 e 8 f 114
g 58 h 5 i 96 j 29 k 300 l 180
m 76 n 59 o 8 p 8

4 a 105 b 45 c 34 d 9 e 125 f 38
g 280 h 80 i 118 j 88 k 104 l 66
m 40 n 72 o 79 p 25

Page 20

1 a 5868 b 7889 c 3959 d 8988 e 6542 f 7845
g 8432 h 5141 i 49 929 j 41 180 k 62 795 l 74 993

2 a 8747 b 7852 c 8546 d 14 982 e 46 733 f 25 795
g 43 168 h 89 810

3 The super answers are:
a 20 189 b 26 754

4 a 82 193 b 107 047 c 74 218 d 122 760 e 84 664 f 79 443

Page 21

1 a 4453 b 2241 c 3132 d 3252 e 1781 f 2075
g 1618 h 2739 i 7142 j 12 085 k 12 874 l 18 335

2 a 4246 b 6289 c 7516 d 5416 e 24 906 f 18 547
g 46 311 h 38 678 i 68 477 j 38 165 k 17 321 l 27 188

3 a 18 938 and 4836; 14 102 b 18 938 and 12 775
c 15 086 and 8032 d 12 775 and 11 941
e 15 086 and 12 775; 11 941 and 9817; 12 775 and 9817

4 a K1851 b K6492

Page 22

1 a 6616 b 15 018 c 5478 d 9346 e 11 024
f 9835 g 10 572 h 20 056 i 6835 j 15 428
k 12 075 l 21 960 m 6447 n 20 531

2 a 13 536 b 16 152 c 9215 d 17 275 e 100 096
f 243 610 g 228 514 h 778 820

3 a 68 384 b 48 464 c 66 082 d 94 809 e 148 654
f 223 740 g 52 139 h 86 432 i 97 728 j 167 313
k 234 468 l 120 846
The name of the town is KIETA.

Page 23

1 a $536\frac{3}{5}$ b 1878 c $815\frac{1}{7}$ d $1532\frac{1}{2}$
e $584\frac{1}{8}$ f $614\frac{5}{6}$ g 1703 h $731\frac{3}{4}$
i $6719\frac{1}{6}$ j $7461\frac{1}{2}$ k $8505\frac{1}{3}$ l 5229

2 a 248 b $114\frac{15}{16}$ c $266\frac{19}{21}$ d $146\frac{4}{7}$
e $159\frac{3}{5}$ f $235\frac{2}{29}$ g $2114\frac{16}{17}$ h $1265\frac{19}{22}$
i $1673\frac{1}{25}$

Pages 24–25

1 a 8 b 10 c 2 d 3 e 5 f 2
g 5 h 3 i 1 j 2 k 2 l 7
m 8 n 1 o 2 p 5 q 2 r 2
s 3 t 5 u 3 v 6 w 10 x 8

2 a 8 b 12 c 20 d 8 e 12 f 16
g 30 h 9 i 21 j 9 k 21 l 10
m 15 n 35 o 24 p 20 q 4 r 18
s 14 t 10 u 9 v 4 w 21 x 28

3 a 18 b 4 c 8 d 6 e 20 f 10
g 6 h 6 i 9 j 8

4 a 75 b 80 c 56 d 80 e 875 f 750
g 75 h 80 i 350 j 60 k 300 l 1250
m 90 n 640 o 150 p 1200 q 100 r 160
s 3500 t 560 u 180 v 2000 w 2100 x 114

Pages 26–27

1 a $\frac{1}{4}, \frac{1}{3}, \frac{1}{5}, \frac{1}{2}, \frac{3}{4}$ b $1\frac{1}{2}, 6\frac{2}{3}, 2\frac{3}{4}$
c $\frac{7}{7}, \frac{9}{7}, \frac{12}{5}, \frac{10}{7}, \frac{11}{5}, \frac{3}{3}, \frac{10}{10}$ d $\frac{7}{7}, \frac{3}{3}, \frac{10}{10}$
e $\frac{9}{7}, \frac{12}{5}, 1\frac{1}{2}, \frac{10}{7}, \frac{11}{5}, 6\frac{2}{3}, 2\frac{3}{4}$ f $\frac{1}{4}, \frac{1}{3}, \frac{1}{5}, \frac{1}{2}, \frac{3}{4}$

2 a $4\frac{1}{2}$ b $3\frac{2}{5}$ c $4\frac{2}{3}$ d $5\frac{1}{4}$ e $2\frac{2}{3}$ f $3\frac{1}{3}$
g $3\frac{1}{6}$ h $3\frac{2}{7}$ i $1\frac{7}{8}$ j $2\frac{6}{7}$ k $3\frac{1}{4}$ l $2\frac{4}{9}$
m $5\frac{1}{3}$ n $1\frac{4}{7}$ o $4\frac{1}{4}$ p $4\frac{1}{6}$ q $4\frac{1}{8}$ r $4\frac{1}{7}$
s $3\frac{4}{9}$ t $1\frac{1}{4}$ u $2\frac{5}{8}$ v $3\frac{2}{11}$ w $1\frac{2}{7}$ x $4\frac{4}{5}$

3 a $\frac{9}{4}$ b $\frac{9}{2}$ c $\frac{7}{4}$ d $\frac{17}{6}$ e $\frac{15}{8}$ f $\frac{33}{10}$
g $\frac{17}{3}$ h $\frac{13}{5}$ i $\frac{35}{8}$ j $\frac{13}{9}$ k $\frac{31}{11}$ l $\frac{7}{5}$
m $\frac{19}{6}$ n $\frac{13}{8}$ o $\frac{13}{2}$ p $\frac{13}{3}$ q $\frac{34}{9}$ r $\frac{20}{7}$
s $\frac{41}{10}$ t $\frac{40}{7}$ u $\frac{25}{11}$ v $\frac{17}{9}$ w $\frac{43}{7}$ x $\frac{19}{5}$

4 a $\frac{5}{3}, \frac{7}{5}, 1\frac{2}{3}$ b $2\frac{1}{7}, \frac{12}{5}, \frac{15}{6}$ c $\frac{9}{10}, \frac{2}{9}, \frac{5}{8}, \frac{7}{12}, \frac{4}{5}$
d $7\frac{1}{4}, 4\frac{3}{4}, \frac{13}{4}, 3\frac{1}{8}$ e $7\frac{1}{4}$

Pages 28–29

1 a $\frac{2}{12}$ b $\frac{3}{9}$ c $\frac{3}{21}$ d $\frac{4}{20}$ e $\frac{4}{6}$ f $\frac{6}{10}$
g $\frac{6}{16}$ h $\frac{12}{16}$ i $\frac{6}{14}$ j $\frac{15}{18}$ k $\frac{6}{27}$ l $\frac{35}{40}$
m $\frac{4}{10}$ n $\frac{20}{35}$ o $\frac{15}{24}$ p $\frac{12}{18}$ q $\frac{14}{18}$ r $\frac{12}{40}$
s $\frac{6}{33}$ t $\frac{16}{36}$ u $\frac{14}{49}$ v $\frac{40}{50}$ w $\frac{90}{100}$ x $\frac{32}{44}$

2 a $\frac{1}{3}$ b $\frac{1}{2}$ c $\frac{1}{4}$ d $\frac{2}{3}$ e $\frac{2}{3}$ f $\frac{2}{3}$
g $\frac{3}{10}$ h $\frac{4}{5}$ i $\frac{4}{5}$ j $\frac{5}{12}$ k $\frac{1}{4}$ l $\frac{3}{4}$
m $\frac{7}{10}$ n $\frac{3}{5}$ o $\frac{3}{4}$ p $\frac{1}{3}$ q $\frac{3}{4}$ r $\frac{3}{7}$
s $\frac{2}{5}$ t $\frac{2}{5}$ u $\frac{5}{6}$ v $\frac{2}{11}$ w $\frac{3}{7}$ x $\frac{2}{5}$

3 a $\frac{2}{8}$ b $\frac{5}{10}$ c $\frac{2}{10}$ d $\frac{1}{2}$ e $\frac{3}{9}$ f $\frac{4}{6}$
g $\frac{10}{100}$ h $\frac{6}{8}$ i $\frac{1}{2}$ j $\frac{2}{12}$

4 a $\frac{5}{10}, \frac{6}{12}$ b $\frac{4}{6}$ c $\frac{6}{12}, \frac{8}{16}$ d $\frac{3}{9}, \frac{2}{6}$
e $\frac{6}{8}$ f $\frac{4}{5}$ g $\frac{1}{4}, \frac{2}{8}, \frac{3}{12}$ h $\frac{1}{5}$
i $\frac{10}{40}$ j $\frac{4}{4}, \frac{9}{9}$

5 a $\neq$ b $\neq$ c = d = e = f $\neq$
g = h $\neq$ i $\neq$ j = k $\neq$ l =
m = n $\neq$ o = p $\neq$ q = r =
s = t $\neq$ u = v $\neq$ w = x =

Pages 30–31

1 a $\frac{5}{6}$ b $\frac{5}{8}$ c $\frac{6}{7}$ d $\frac{3}{10}$ e $\frac{3}{4}$ f $\frac{7}{12}$
g $\frac{9}{10}$ h $\frac{5}{12}$ i $\frac{2}{5}$ j $\frac{7}{8}$ k $1\frac{2}{3}$ l $1\frac{3}{10}$
m $1\frac{1}{5}$ n $\frac{6}{11}$ o $\frac{2}{15}$ p $1\frac{1}{2}$ q $\frac{4}{7}$ r $1\frac{2}{9}$
s $\frac{9}{25}$ t $1\frac{1}{2}$ u $1\frac{2}{5}$ v $\frac{3}{8}$ w $1\frac{2}{9}$ x $\frac{8}{15}$

2 a $1\frac{1}{4}$ b $1\frac{1}{9}$ c $\frac{3}{4}$ d $\frac{1}{2}$ e $\frac{11}{20}$ f $\frac{2}{15}$
g $1\frac{5}{8}$ h $1\frac{1}{8}$ i $1\frac{7}{16}$ j $1\frac{1}{12}$ k $\frac{1}{2}$ l $\frac{1}{4}$
m $\frac{8}{15}$ n $\frac{1}{25}$ o $\frac{7}{12}$ p $1\frac{4}{9}$ q $1\frac{1}{3}$ r $1\frac{1}{5}$
s $\frac{3}{16}$ t $\frac{4}{21}$ u $\frac{9}{22}$ v $\frac{7}{15}$ w $1\frac{2}{25}$ x $\frac{23}{24}$

3 a 1 b $\frac{5}{8}$ c $\frac{7}{10}$ d $1\frac{3}{6}$ e $1\frac{1}{8}$ f $\frac{5}{8}$
g $\frac{2}{10}$ h 1 i $\frac{7}{12}$ j $\frac{2}{9}$ k $\frac{1}{4}$ l $1\frac{3}{4}$
m $1\frac{3}{16}$ n $\frac{4}{15}$

4 a

+	$\frac{1}{4}$	$\frac{1}{2}$	$\frac{1}{8}$	$\frac{3}{4}$	$\frac{3}{8}$	1
$\frac{1}{4}$	$\frac{1}{2}$	$\frac{3}{4}$	$\frac{3}{8}$	1	$\frac{5}{8}$	$1\frac{1}{4}$
$\frac{1}{2}$	$\frac{3}{4}$	1	$\frac{5}{8}$	$1\frac{1}{4}$	$\frac{7}{8}$	$1\frac{1}{2}$

b

+	$\frac{1}{2}$	1	$\frac{3}{4}$	$\frac{7}{8}$	2	$\frac{5}{8}$
$\frac{1}{2}$	1	$1\frac{1}{2}$	$1\frac{1}{4}$	$1\frac{3}{8}$	$2\frac{1}{2}$	$1\frac{1}{8}$
$\frac{1}{4}$	$\frac{3}{4}$	$1\frac{1}{4}$	1	$1\frac{1}{8}$	$2\frac{1}{4}$	$\frac{7}{8}$

5 a $\frac{2}{3}$ b $\frac{1}{6}$ c $\frac{5}{8}$ d $1\frac{1}{10}$ e $\frac{3}{4}$ f $\frac{1}{8}$
g $\frac{5}{8}$ h $\frac{5}{12}$ i $\frac{2}{5}$ j $\frac{1}{4}$ k $\frac{1}{2}$ l $\frac{5}{9}$
m $3\frac{1}{3}$ n $3\frac{3}{7}$ o $2\frac{2}{5}$ p $4\frac{5}{6}$ q $\frac{1}{6}$ r $\frac{1}{6}$
s $\frac{2}{5}$ t $\frac{3}{4}$ u $\frac{8}{9}$ v $\frac{3}{8}$ w $\frac{1}{3}$ x $3\frac{3}{8}$

Page 32

1 $4\frac{3}{8}$ **2** $2\frac{5}{8}$ **3** $6\frac{1}{3}$ **4** $3\frac{1}{2}$ **5** $4\frac{1}{9}$ **6** $1\frac{1}{8}$
7 $5\frac{7}{12}$ **8** $3\frac{3}{10}$ **9** $7\frac{7}{8}$ **10** $3\frac{3}{8}$ **11** $5\frac{4}{9}$ **12** $2\frac{1}{6}$
13 $6\frac{7}{10}$ **14** $4\frac{5}{9}$

Page 33

1 a 8.63 b 7.15 c 83.4 d 92.06 e 12.7 f 80.4
g 56.4 h 3.19 i 203.06 j 8.47

2 a 6 b $\frac{9}{10}$ c 50 d $\frac{6}{10}$ e $\frac{5}{100}$ f $\frac{4}{10}$
g $\frac{7}{100}$ h $\frac{0}{10}$ i 200 j $\frac{7}{10}$ k $\frac{8}{10}$ l $\frac{6}{100}$
m $\frac{2}{1000}$ n 9 o $\frac{8}{100}$ p $\frac{6}{1000}$ q $\frac{9}{10}$ r 5

3 a 56.7 b 35.71 c 4.57 d 56.34

4 a 596.18 b 79.8 c 5.208 d 4.018

5 a $(2 \times 10) + (5 \times 1) + (8 \times \frac{1}{10})$ b $(3 \times 1) + (6 \times \frac{1}{10}) + (7 \times \frac{1}{100})$
c $(1 \times 10) + (2 \times 1) + (3 \times \frac{1}{10})$ d $(4 \times 1) + (3 \times \frac{1}{10}) + (8 \times \frac{1}{1000})$
e $(6 \times 1) + (9 \times \frac{1}{10}) + (3 \times \frac{1}{100})$ f $(2 \times 1) + (1 \times \frac{1}{100}) + (5 \times \frac{1}{1000})$

6 a add 0.1 b subtract 0.2 c add 2 d subtract 0.03
e add 0.2 f add 0.3 g subtract 0.04 h subtract 0.02
i add 0.03 j add 0.003 k add 2 l subtract 0.2

Page 34

1 a 3.74 b 12.8 c 4.15 d 59 e 12.26 f 8.08
g 14.45 h 7.11 i 6.05 j 36.5 k 82.46 l 4.178
m 2.462 n 3.91 o 26.3 p 4.689 q 6.17 r 32.02
s 6.09 t 18 u 2.195 v 40.812 w 4.017 x 76.9

2 a 3.7 b 6.95 c 27.9 d 14 e 6.62 f 51.5
g 8.029 h 2.99 i 46.9 j 6.052 k 4.25 l 18.92
m 4.933 n 6.04 o 13 p 52.538 q 7.71 r 6.801
s 7.9 t 2.91 u 5.816 v 7.93 w 16.077 x 4.88

3 a 4.87 b 3.162 c 17.1 d 6.405 e 27.01 f 5.013
g 2.57 h 87.21 i 3.97 j 9.208 k 41.073 l 58.01
m 1.04 n 51.77 o 2.093 p 8.71 q 16.6 r 70.81
s 6 t 4.713 u 15.71 v 33.42 w 18.01 x 25.17

4 a 7.82 b 5.95 c 15.8 d 3.66 e 39.79 f 63.692
g 19.03 h 24.991 i 35.99 j 5.71 k 19.29 l 64.99
m 9.005 n 11.59 o 20.07 p 8.702 q 6.063 r 12.79
s 48.99 t 5.01 u 6.53 v 8.127 w 3.68 x 17.49

5 a 1.638 b 8.046 c 2.611 d 15.088 e 23.001 f 5.571
g 45.091 h 32.801 i 55.1 j 7.401 k 29.22 l 13.601
m 2.973 n 35.01 o 85.881 p 21.061 q 45.09 r 3.121
s 5.801 t 64.001 u 37.601 v 1.41 w 81.201 x 3.68

6 a 5.812 b 2.654 c 17.84 d 16.999 e 22.01 f 9.819
g 33.093 h 57.809 i 14.089 j 3.799 k 48.699 l 23.499
m 11.069 n 61.109 o 32.039 p 5.62 q 59.999 r 9.499
s 28.599 t 3 u 67.119 v 83.799 w 7.639 x 51.999

Page 35

1 a 0.3 b 0.9 c 0.63 d 0.19 e 0.03 f 0.27
g 0.654 h 0.123 i 0.065 j 0.37 k 0.1 l 0.05
m 0.71 n 0.009 o 0.6 p 0.4

2 a 1.6 b 7.09 c 3.45 d 17.125 e 10.41 f 25.75
g 30.04 h 9.087 i 6.14 j 47.3 k 52.011 l 8.16
m 2.29 n 31.9 o 4.06

3 a $\frac{4}{5}$ b $6\frac{3}{4}$ c $1\frac{317}{1000}$ d $\frac{87}{1000}$ e $8\frac{1}{200}$ f $2\frac{19}{100}$
g $5\frac{3}{100}$ h $17\frac{4}{25}$ i $24\frac{13}{250}$ j $31\frac{4}{5}$ k $12\frac{3}{50}$ l $7\frac{509}{1000}$
m $1\frac{23}{1000}$ n $\frac{67}{100}$ o $3\frac{411}{1000}$ p $9\frac{3}{25}$ q $18\frac{3}{4}$ r $73\frac{3}{5}$

4 a $\frac{1}{4}$ b $\frac{3}{4}$ c $\frac{1}{2}$ d $\frac{7}{100}$ e $\frac{3}{100}$ f $\frac{37}{100}$
g $\frac{39}{50}$ h $\frac{3}{20}$ i $\frac{61}{100}$ j $\frac{1}{10}$ k $\frac{1}{5}$ l $\frac{9}{10}$
m $\frac{39}{100}$ n $\frac{1}{20}$ o $\frac{21}{25}$ p $\frac{19}{100}$ q $\frac{9}{20}$ r $\frac{11}{100}$
s $\frac{9}{100}$ t $\frac{91}{100}$

5 a 0.57, 57% b $\frac{21}{100}$, 0.21 c $\frac{39}{100}$, 39% d $\frac{7}{10}$, 70%
e $\frac{9}{100}$, 0.09 f 0.86, 86% g 0.3, 30% h $\frac{2}{100}$, 2%

Page 36

1 a 0.85 b 0.7 c 0.4 d 1.5 e 2.9 f 5.2
g 0.96 h 1.85 i 2.5 j 0.58 k 6.25 l 3.2
m 1.675 n 5.08 o 4.7 p 2.5 q 8.062 r 7.11
s 2.4 t 3.05

2 a 1.65 b 5.72 c 9.86 d 7.3 e 3.6 f 2.75
g 14.06 h 9.3 i 6.3 j 8.7 k 4.67 l 0.6
m 1.24 n 3.64 o 2.811

3 a $\frac{2}{10}$ b 0.6 c $\frac{9}{100}$ d $\frac{3}{10}$ e 5% f $\frac{75}{100}$
g 20% h 0.72 i 0.49 j 40% k 0.03 l 0.9
m 87% n $\frac{9}{10}$ o $\frac{2}{100}$

4 a 58%, 0.6, $\frac{61}{100}$, 65% b 0.08, 0.75, $\frac{8}{10}$, 81%
c $\frac{4}{10}$, 0.48, 49%, $\frac{1}{2}$ d 0.02, 19%, $\frac{2}{10}$, 0.25
e $\frac{25}{100}$, 0.3, 31%, 33% f 7%, 0.7, 0.72, 75%
g 0.08, 0.75, $\frac{8}{10}$, 85% h 0.7, 72%, $\frac{75}{100}$, 79%
i 3%, 0.3, 0.315, $\frac{35}{100}$ j $\frac{2}{100}$, 21%. 0.225, 2.1
k 0.015, $\frac{10}{100}$, 15%, 1.5 l 0.053, 0.5, $\frac{51}{100}$, 53%

Page 37

1 a 8 kg b 9 kg c 2 kg d 1 kg e 5 kg f 12 kg
g 3 kg h 13 kg i 16 kg j 10 kg k 9 kg l 14 kg
m 3 kg n 6 kg o 1 kg p 4 kg q 1 kg r 4 kg
s 1 kg t 7 kg

2 a 9 km b 6 km c 13 km d 23 km e 16 km f 22 km
g 19 km h 45 km i 31 km j 27 km k 35 km l 28 km
m 8 km n 17 km o 27 km p 11 km q 18 km r 33 km
s 46 km t 19 km

3 a K5 b K8 c K3 d K6 e K6 f K4
g K4 h K2 i K5 j K5 k K10 l K11
m K16 n K10 o K18 p K17 q K9 r K8

4 a 4.2 L b 7.6 L c 1.4 L d 0.9 L e 10.7 L f 15.8 L
g 6.1 L h 8.3 L i 12.4 L j 9.3 L k 5.9 L l 2.7 L
m 11.4 L n 17.9 L o 8.1 L p 19.1 L q 7 L r 4.9 L

5 a 11 s, 10.9 s b 19 s, 18.9 s c 9 s, 9.3 s d 12 s, 11.8 s
e 11 s, 11 s f 8 s, 8.5 s g 12 s, 12.3 s h 12 s, 11.6 s
i 9 s, 8.8 s j 9 s, 9 s k 10 s, 10.5 s l 13 s, 12.9 s
m 6 s, 5.9 s n 16 s, 15.6 s o 25 s, 25.2 s p 18 s, 18.1 s
q 25 s, 24.7 s r 21 s, 21.3 s

6 a 3, 2.5, 2.55 b 2, 1.8, 1.75 c 5, 5.1, 5.08 d 8, 8.2, 8.25
e 1, 0.6, 0.63 f 4, 3.8, 3.78 g 9, 9.4, 9.4 h 6, 6.2, 6.19
i 7, 7.4, 7.35 j 5, 4.9, 4.89

Page 38

1 a 13.4 b 16 c 14.7 d 14.9 e 39.4 f 62.2
g 2.2 h 2.7 i 3.7 j 6.7 k 6.8 l 7.8
m 14.46 n 15.65 o 43.85 p 44.2 q 16.768 r 14.722

2 a 24.1 b 64.47 c 25.727 d 50.96 e 72.207 f 27.412
g 41.04 h 42.65 i 1.294 j 7.32 k 8.535 l 12.663

Page 39

1 a 31.5 b 17.1 c 16.8 d 29.6 e 47.6 f 20.64
g 11.67 h 13.75 i 39.41 j 18.32

2 a 197.76 b 192.54 c 917.5 d 470.1 e 191.8 f 12.268
g 68.715 h 40.915 i 114.064 j 303.232 k 1043.2 l 53.964
m 30.66 n 687.42 o 45.78 p 437.4 q 139.24 r 35.296
s 3537.6 t 208.95 u 11.007 v 1092.49 w 266.4 x 32.6

3 a K63.30 b K73.50 c K27.92 d K34.75 e K50.60 f K10.44

Page 40

1 a 3.4 b 3.2 c 6.2 d 3.1 e 4.1 f 2.21
g 2.32 h 7.1 i 10.1 j 13.22 k 24.13 l 3.11

2 a 1.4 b 2.1 c 3.1 d 1.1 e 2.1 f 8.2
g 2.1 h 4.2 i 8.21 j 5.11 k 4.02 l 5.32
m 1.21 n 1.42 o 1.31 p 1.51 q 1.21 r 2.73
s 3.73 t 2.31 u 131.5 v 142.1 w 178.2 x 267.6
y 13.42 z 17.54

Page 41

1 17.5 kg

2 a K16.80 b K39.20 c K33.60 d K44.80

3 a K903.20 b K277.80

4 243.75 km **5** K8.28 **6** 271.8 kg

7 K22.55 **8** 864.75 litres **9** 6428 km

10 62.8 L **11** 83.7 km **12** 6.2 metres

13 K1249.60

Assessment

Page 42

1 b **2** c **3** c **4** b **5** d **6** d

7 a **8** c **9** b **10** a **11** d **12** a

13 b **14** c **15** b **16** d **17** a **18** c

19 b **20** b

Pages 43–44

1 a 47 b 49 c 78 d 51 e 77 f 60

2 a 105 b 137 c 79 d 165 e 154 f 307
g 168 h 189 i 318 j 370

3 a 43 b 52 c 52 d 37 e 77 f 71
g 131 h 109 i 313 j 322

4 a 42 b 32 c 54 d 96 e 56 f 72
g 72 h 35 i 64 j 49

5 a 7 b 5 c 7 d 4 e 5 f 6
g 3 h 9 i 5 j 9

6 a 320 b 180 c 350 d 360 e 560 f 70
g 40 h 90 i 60 j 40 k 600 l 1500
m 2000 n 1200 o 1800 p 9 q 9 r 7
s 7 t 7

7 a 130 b 110 c 110 d 170 e 120 f 70
g 30 h 50 i 60 j 170 k 230 l 210
m 270 n 300 o 340 p 80 q 170 r 280
s 120 t 240

8 a 216 b 358 c 796 d 100 e 899 f 534
g 5 h 421 i 2807 j 234 k 477 l 16

9 a 28 b 100 c 80 d 10 e 185 f 12
g 171 h 64

10 a 7999 b 6889 c 8635 d 29 457 e 65 482 f 52 163
g 58 254 h 28 901 i 70 734 j 38 696 k 51 405 l 60 845

11 a 3523 b 5223 c 3514 d 2291 e 2217 f 2460
g 13 906 h 17 455 i 10 483 j 43 462 k 55 745 l 40 847

12 a 6948 b 10 470 c 15 410 d 11 032 e 32 249
f 68 800 g 194 064 h 35 904 i 195 642 j 103 772

13 a 96 404 b 65 765 c 61 574 d 121 584 e 78 242

14 a $760\frac{1}{7}$ b $927\frac{2}{5}$ c $1835\frac{1}{4}$ d $995\frac{1}{3}$
e $583\frac{5}{8}$ f $2035\frac{2}{3}$ g $1340\frac{1}{3}$ h $370\frac{4}{7}$
i $944\frac{1}{2}$ j $336\frac{3}{5}$ k 2091 l 7683
m $4313\frac{1}{5}$ n $2814\frac{5}{7}$ o 8956

15 a 405 b $191\frac{18}{25}$ c $222\frac{26}{31}$ d $260\frac{9}{22}$
e $298\frac{13}{18}$ f $204\frac{1}{36}$ g $289\frac{27}{29}$ h $182\frac{5}{21}$
i $1571\frac{9}{23}$ j $2790\frac{5}{19}$ k $2562\frac{1}{4}$ l $1799\frac{23}{27}$

Pages 45–46

1 a 12 b 4 c 11 d 8 e 3 f 7
g 3 h 8 i 2 j 9 k 21 l 10
m 24 n 35 o 15 p 15 q 20 r 27
s 24 t 15

2 a 12 b 6 c 4 d 18 e 16

3 a $2\frac{3}{7}$ b $2\frac{2}{9}$ c $3\frac{3}{4}$ d $4\frac{1}{6}$ e $6\frac{1}{5}$ f $6\frac{1}{3}$
g $4\frac{2}{3}$ h $4\frac{3}{8}$ i $6\frac{2}{3}$ j $9\frac{1}{4}$ k $3\frac{5}{7}$ l $5\frac{3}{4}$
m $2\frac{5}{8}$ n $8\frac{3}{5}$ o $3\frac{1}{7}$ p $5\frac{3}{10}$ q $3\frac{2}{9}$ r $2\frac{1}{6}$

4 a $\frac{23}{8}$ b $\frac{11}{6}$ c $\frac{26}{7}$ d $\frac{45}{8}$ e $\frac{15}{4}$ f $\frac{14}{3}$
g $\frac{9}{5}$ h $\frac{59}{10}$ i $\frac{23}{5}$ j $\frac{13}{2}$ k $\frac{19}{8}$ l $\frac{37}{5}$
m $\frac{27}{8}$ n $\frac{14}{9}$ o $\frac{17}{3}$ p $\frac{26}{9}$ q $\frac{37}{4}$ r $\frac{31}{7}$

5 a $\frac{8}{12}$ b $\frac{8}{9}$ c $\frac{15}{25}$ d $\frac{2}{5}$ e $\frac{8}{14}$ f $\frac{3}{5}$
g $\frac{20}{32}$ h $\frac{3}{4}$ i $\frac{12}{32}$ j $\frac{3}{5}$ k $\frac{42}{54}$ l $\frac{3}{4}$

6 a $1\frac{2}{9}$ b $\frac{2}{15}$ c $1\frac{3}{5}$ d $\frac{5}{12}$ e $1\frac{1}{4}$ f $\frac{3}{10}$
g $1\frac{5}{8}$ h $\frac{2}{15}$ i $1\frac{2}{9}$ j $\frac{1}{6}$ k $1\frac{5}{12}$ l $\frac{3}{8}$

7 a $6\frac{3}{8}$ b $3\frac{1}{10}$ c $6\frac{1}{2}$ d $3\frac{1}{3}$ e $7\frac{8}{9}$ f $3\frac{5}{12}$
g $8\frac{3}{8}$ h $4\frac{1}{4}$ i $6\frac{1}{2}$ j $3\frac{3}{8}$ k $5\frac{5}{6}$ l $3\frac{5}{14}$

Pages 47–48

1 a 7 b $\frac{5}{10}$ c $\frac{8}{100}$ d $\frac{7}{10}$ e $\frac{9}{1000}$ f $\frac{2}{10}$
g $\frac{4}{100}$ h 5 i 90 j $\frac{4}{10}$ k $\frac{9}{100}$ l $\frac{4}{1000}$

2 a 52.48 b 9.147 c 63.052 d 6.289

3 a 0.73 b 15.9 c 27.752 d 8.16 e 5.05 f 13

4 a 5.82 b 22.59 c 36.064 d 9.1 e 23.99 f 0.095

5 a 0.7 b 0.39 c 0.03 d 0.147 e 0.078 f 0.009
g 2.5 h 8.27 i 12.8 j 6.01 k 5.083 l 10.069
m 2.19 n 15.008 o 26.3

6 a $\frac{3}{5}$ b $3\frac{47}{100}$ c $1\frac{9}{100}$ d $6\frac{751}{1000}$ e $2\frac{37}{1000}$ f $11\frac{2}{5}$

7 a $\frac{1}{2}$ b $\frac{3}{10}$ c $\frac{9}{100}$ d $\frac{23}{100}$ e $\frac{3}{50}$ f $\frac{11}{20}$

8 a 0.75 b 0.03 c 0.56 d 0.81 e 0.08 f 0.49

9 a 0.8, 79%, 0.079, $\frac{7}{100}$ b 25%, 0.23, $\frac{2}{10}$, 0.02
c 35%, 0.33, $\frac{30}{100}$, $\frac{3}{100}$ d 5.8, 58%, $\frac{5}{10}$, 0.058
e 87%, 0.8, $\frac{87}{1000}$, 0.08 f 1.5, 0.15, 10%, $\frac{15}{1000}$

10 a 7.4 b 4.1 c 2.5 d 9.1 e 5.3 f 8.7

11 a 1.37 b 0.82 c 5.03 d 8.21 e 6.05 f 2.4

12 a 16.1 b 3.2 c 25.1 d 2.7 e 52.5 f 7.5
g 52.14 h 14.08 i 12.947 j 3.187 k 525.39 l 26.35

13 a 23.2 b 60.9 c 19 d 22.89 e 15.42 f 184.64
g 155.35 h 26.968 i 72.464 j 4634.1 k 1663.2 l 249.84

14 a 2.3 b 3.1 c 3.2 d 6.1 e 1.41 f 1.91
g 1.72 h 2.54 i 230.9 j 21.8 k 193.8 l 46.6

15 a K536.65 b 12.32 metres c 167.55 kg

Strand: Measurement

Page 49

1 a 65 mm, 6 cm 5 mm, 6.5 cm b 40 mm, 4 cm, 4 cm
c 100 mm, 10 cm, 10 cm d 75 mm, 7 cm 5 mm, 7.5 cm
e 35 mm, 3 cm 5 mm, 3.5 cm f 45 mm, 4 cm 5 mm, 4.5 cm
g 70 mm, 7 cm, 7 cm h 80 mm, 8 cm, 8 cm
i 25 mm, 2 cm 5 mm, 2.5 cm j 130 mm, 13 cm, 13 cm
k 15 mm, 1 cm 5 mm, 1.5 cm l 85 mm, 8 cm 5 mm, 8.5 cm
m 120 mm, 12 cm, 12 cm n 20 mm, 2 cm, 2 cm
o 55 mm, 5 cm 5 mm, 5.5 cm p 90 mm, 9 cm, 9 cm
q 42 mm, 4 cm 2 mm, 4.2 cm r 48 mm, 4 cm 8 mm, 4.8 cm

2 a – h Teacher to check.

Page 50

1 a 200 cm b 500 cm c 100 cm d 400 cm e 1000 cm
f 1200 cm g 350 cm h 650 cm i 150 cm j 950 cm

2 a 7 m b 3 m c 8 m d 11 m e 6 m
f $7\frac{1}{2}$ m g $2\frac{1}{2}$ m h $4\frac{1}{2}$ m i $12\frac{1}{2}$ m j $15\frac{1}{2}$ m

3 a 80 mm b 50 mm c 110 mm d 30 mm e 140 mm
f 25 mm g 85 mm h 105 mm i 45 mm j 75 mm

4 a 5 cm b 1 cm c 10 cm d 7 cm e 12 cm
f $6\frac{1}{2}$ cm g $11\frac{1}{2}$ cm h $2\frac{1}{2}$ cm i $9\frac{1}{2}$ cm j $1\frac{1}{2}$ cm

5 a 2000 m b 5000 m c 9000 m d 4000 m e 10 000 m
f 5500 m g 500 m h 11 500 m i 6500 m j 3500 m

6 a 1 km b 6 km c 3 km d 8 km e 13 km
f $2\frac{1}{2}$ km g $8\frac{1}{2}$ km h $16\frac{1}{2}$ km i $1\frac{1}{2}$ km j $10\frac{1}{2}$ km

7 a 1 m 23 cm b 2 m 47 cm c 1 m 8 cm d 3 m 61 cm
e 2 m 96 cm f 13 m 14 cm g 25 m 59 cm h 12 m 82 cm
i 34 m 5 cm j 51 m 17 cm

8 a 8 cm 9 mm b 1 cm 6 mm c 6 cm 3 mm d 12 cm 5 mm
e 27 cm 1 mm f 10 cm 3 mm g 33 cm 4 mm h 21 cm 8 mm
i 56 cm 6 mm j 48 cm 5 mm

9 a 1 km 700 m b 1 km 345 m c 3 km 678 m d 2 km 97 m
e 1 km 820 m f 14 km 562 m g 10 km 204 m h 47 km 35 m
i 32 km 712 m j 26 km 955 m

Page 51

1 12 cm **2** 11 cm **3** 14 cm **4** 14 cm **5** 21 cm

6 22 cm **7** 23 cm **8** 21 cm **9** 14 cm **10** 22 cm

11 11 cm **12** 18 cm

Page 52

1 8 cm^2 **2** 9 cm^2 **3** 10 cm^2 **4** 11 cm^2 **5** 16 cm^2

6 9 cm^2 **7** 8 cm^2 **8** 12 cm^2 **9** $8\frac{1}{2}$ cm^2 **10** 10 cm^2

11 10 cm^2

Page 53

1 1. 6 cm, 2 cm, 12 cm^2 **2** 4 cm, 2 cm, 8 cm^2 **3** 5 cm, 4 cm, 20 cm^2

4 5 cm, 3 cm, 15 cm^2 **5** 4 cm, 3 cm, 12 cm^2 **6** 5 cm, 2 cm, 10 cm^2

7 4 cm, 4 cm, 16 cm^2 **8** 3 cm, 2 cm, 6 cm^2 **9** 7 cm, 4 cm, 28 cm^2

10 3 cm, 3 cm, 9 cm^2 **11** 9 cm, 3 cm, 27 cm^2 **12** 5 cm, 5 cm, 25 cm^2

Page 54

1 12 m^2 **2** 6 m^2 **3** 10.5 m^2 **4** 12.5 m^2

5 17.5 m^2 **6** 14 m^2 **7** 16.5 m^2 **8** 9 m^2

9 22.5 m^2 **10** 13.5 m^2 **11** 28.5 m^2 **12** 7.5 m^2

Page 55

1 **a** 1 L 540 mL **b** 2 L 735 mL **c** 5 L 61 mL **d** 1 L 7 mL
e 3 L 604 mL **f** 7 L 10 mL **g** 4 L 218 mL **h** 3 L 720 mL
i 2 L 906 mL **j** 1 L 700 mL **k** 6 L 194 mL **l** 2 L 733 mL
m 4 L 372 mL **n** 5 L 995 mL **o** 2 L 36 mL **p** 1 L 877 mL

2 **a** 8000 mL **b** 3000 mL **c** 5500 mL **d** 2500 mL
e 6070 mL **f** 8510 mL **g** 1015 mL **h** 3407 mL
i 250 mL **j** 4250 mL **k** 9500 mL **l** 500 mL
m 7326 mL **n** 2750 mL **o** 5600 mL **p** 532 mL
q 4900 mL **r** 3150 mL **s** 1700 mL **t** 8250 mL

3 **a** 3.5 L **b** 5.25 L **c** 2.75 L **d** 0.5 L **e** 2.765 L **f** 4.86 L
g 1.485 L **h** 3.038 L **i** 0.25 L **j** 0.9 L **k** 5.002 L **l** 4.8 L

4 **a** 6900 mL **b** 7675 mL **c** 7350 mL **d** 3710 mL
e 5990 mL **f** 7788 mL **g** 13 105 mL **h** 4450 mL
i 7005 mL **j** 3166 mL

5 **a** 2 m^3 **b** 5 m^3 **c** 8 m^3 **d** 3 m^3 **e** 1.5 m^3 **f** 4.5 m^3
g 10 m^3 **h** 6.5 m^3 **i** 3.25 m^3 **j** 9.75 m^3 **k** 2.25 m^3 **l** 4.6 m^3
m 1.3 m^3 **n** 5.7 m^3 **o** 8.1 m^3 **p** 3.8 m^3

6 **a** 500 000 cm^3 **b** 900 000 cm^3 **c** 600 000 cm^3 **d** 250 000 cm^3
e 750 000 cm^3 **f** 125 000 cm^3 **g** 425 000 cm^3 **h** 575 000 cm^3
i 50 000 cm^3 **j** 1 075 000 cm^3 **k** 260 000 cm^3 **l** 590 000 cm^3
m 1 150 000 cm^3 **n** 130 000 cm^3 **o** 1 225 000 cm^3 **p** 340 000 cm^3
q 230 000 cm^3 **r** 320 000 cm^3 **s** 170 000 cm^3 **t** 560 000 cm^3

Page 56

1 A = 11 m^3, B = 6 m^3, C = 8 m^3, D = 12 m^3, E = 14 m^3, F = 10 m^3, G = 10 m^3, H = 13 m^3, I = 10 m^3

2 E **3** B

4 **a** 7 m^3 **b** 2 m^3 **c** 1 m^3 **d** 5 m^3

5 **a** 24 m^3 **b** 23 m^3 **c** 4 m^3 **d** 2 m^3

Page 57

1 4 cm, 2 cm, 2 cm, 16 cm^3; **2** 5 cm, 2 cm, 1 cm, 10 cm^3;

3 4 cm, 2 cm, 3 cm, 24 cm^3; **4** 4 cm, 3 cm, 2 cm, 24 cm^3;

5 3 cm, 1 cm, 3 cm, 9 cm^3; **6** 6 cm, 2 cm, 3 cm, 36 cm^3;

7 4 cm, 4 cm, 2 cm, 32 cm^3; **8** 8 cm, 4 cm, 2 cm, 64 cm^3

Pages 58–59

1 675 g, 1.2 kg, 0.86 kg, 0.615 kg, 1010 g, 1.7 kg, 4.1 kg, 725 g, 510 g, 1250 g

2 0.09 kg, 240 g, 100 g, 0.075 kg, 0.05 kg, 106 g, 0.136 kg, 87 g, 0.175 kg

3 0.5 kg, 615 g, 450 g, 0.312 kg, 508 g, 0.74 kg, 0.607 kg

4 0.78 kg, 1240 g, 0.98 kg, 1.02 kg, 1.065 kg, 0.95 kg, 1.23 kg

5 650 kg, 0.35 t, 0.4 t, 740 kg, 0.25 t, 0.7 t, 400 kg, 347 kg

6 1350 kg, 1.25 t, 820 kg, 0.9 t, 1275 kg, 0.85 t, 815 kg, 0.91 t, 1.45 t, 1000 kg

7 **a** 1690 g, 6150 g, $6\frac{1}{4}$ kg, 6.5 kg **b** 0.035 t, 275 kg, 0.3 t, 305 kg
c 0.75 t, 7.03 t, 7500 kg, 7.55 t **d** 1300 g, 1.45 kg, $1\frac{1}{2}$ kg, 1600 g
e 0.053 kg, 0.5 kg, 0.53 kg, 534 g **f** 4.01 t, 4050 kg, 4100 kg, 4.15 t
g 806 kg, 0.87 t, 8.06 t, 8.6 t **h** 230 g, 2030 g, 2.13 kg, 2.3 kg
i 0.049 kg, 0.49 kg, 493 g, 4.9 kg **j** 1.065 t, 1.16 t, 1550 kg, 1.6 t
k 0.9 t, 0.95 t, 8900 kg, 9 t **l** 1.03 kg, 1285 g, 1300 g, 1.34 kg
m 5.025 kg, 5.2 kg, $5\frac{1}{4}$ kg, 5270 g **n** 0.075 t, 0.7 t, 715 kg, 0.75 t
o 255 kg, 2.15 t, $2\frac{1}{2}$ t, 2550 kg **p** 8.09 kg, 8100 g, 8.17 kg, $8\frac{1}{4}$ kg
q 1.075 kg, 1.7 kg, $1\frac{3}{4}$ kg, 1785 g **r** 6070 kg, 6.075 t, 6.17 t, 6.7 t

Page 60

1 **a** 500 g **b** 200 g **c** 2500 g **d** 250 g **e** 750 g **f** 100 g
g 700 g **h** 3200 g **i** 125 g **j** 2800 g **k** 1750 g **l** 4250 g
m 4750 g **n** 2400 g **o** 8500 g **p** 1250 g **q** 800 g **r** 375 g

2 a 750 kg b 3500 kg c 2250 kg d 875 kg e 1900 kg f 10 500 kg g 3375 kg h 600 kg i 1625 kg j 5875 kg k 6300 kg l 11 250 kg m 12 375 kg n 10 200 kg o 9875 kg p 11 800 kg q 15 500 kg r 12 750 kg

3 a $4\frac{1}{2}$ kg b $7\frac{1}{2}$ kg c $2\frac{1}{4}$ kg d $8\frac{1}{4}$ kg e $2\frac{3}{4}$ kg f $6\frac{1}{4}$ kg g $5\frac{1}{10}$ kg h $4\frac{1}{8}$ kg i $7\frac{9}{10}$ kg j $6\frac{1}{8}$ kg

4 a 2500 kg b 7500 kg c 4250 kg d 9750 kg e 250 kg f 1750 kg g 5600 kg h 8300 kg i 900 kg j 10 200 kg k 4800 kg l 12 400 kg m 6125 kg n 3078 kg o 193 kg p 672 kg q 6934 kg r 5007 kg

5 a 2.3 kg b 4.1 kg c 1.6 kg d 0.85 kg e 2.63 kg f 1.875 kg g 3.56 kg h 0.795 kg i 4.238 kg j 3.08 kg k 2.704 kg l 4.16 kg m 0.425 kg n 1.036 kg o 8.01 kg p 7.3 kg q 0.239 kg r 4.129 kg

6 a $1\frac{1}{2}$ t b $10\frac{1}{2}$ t c $6\frac{1}{4}$ t d $3\frac{1}{4}$ t e $11\frac{3}{4}$ t f $9\frac{3}{4}$ t g $12\frac{1}{10}$ t h $15\frac{9}{10}$ t i $10\frac{1}{5}$ t j $9\frac{1}{8}$ t

7 a 600 g b 1400 g c 6500 g d 2700 g e 9100 g f 3800 g g 11 250 g h 7250 g i 12 750 g j 20 650 g k 10 420 g l 8170 g m 156 g n 1907 g o 5077 g p 11 040 g q 15 030 g r 9163 g

Page 61

1 a 300 s b 600 s c 240 s d 150 s e 495 s f 225 s g 390 s h 80 s i 720 s j 1800 s

2 a 180 min b 420 min c 210 min d 135 min e 320 min f 105 min g 630 min h 252 min i 100 min j 45 min

3 a 48 h b 120 h c 240 h d 168 h e 84 h f 204 h g 150 h h 30 h i 114 h j 234 h

4 a 3 hours 5 min b 1 hour 15 min c 3 hours 50 min d 5 hours 10 min e 2 hours 25 min f 1 hour 30 min g 4 hours 13 min h 3 hours 11 min i 6 hours 12 min j 8 hours 20 min

5 a 1 day 6 hours b 2 days 2 hours c 10 days 10 hours d 1 day 3 hours e 1 day 23 hours f 3 days 3 hours g 1 day 9 hours h 2 days 7 hours i 1 day 1 hour j 4 days 4 hours

6 a 1 week 3 days b 2 weeks 6 days c 2 weeks 2 days d 5 weeks 5 days e 3 weeks 4 days f 4 weeks 2 days g 10 weeks 2 days h 7 weeks 1 day i 5 weeks 1 day j 6 weeks 3 days

7 a 48 b 90 c 80 d 150 e 600 f 275

Page 62

1 $\frac{1}{2}$ past 2, 2:30 **2** 10 past 7, 7:10 **3** 5 to 4, 3:55

4 $\frac{1}{4}$ past 10, 10:15 **5** $\frac{1}{4}$ to 1, 12:45 **6** 20 past 8, 8:20

7 10 to 3, 2:50 **8** 25 past 9, 9:25 **9** 20 to 4, 3:40

10 5 past 6, 6:05 **11** 25 to 10, 9:35 **12** $\frac{1}{2}$ past 5, 5:30

13 20 to 7, 6:40 **14** 10 past 11, 11:10 **15** $\frac{1}{4}$ to 2, 1:45

Page 63

1 9:20 a.m., 6:45 a.m., 3:50 a.m., 11:20 a.m., 7:30 a.m., 2:10 a.m., 7:05 a.m., 6:30 a.m.

2 a 7:30 a.m., 11 a.m., 7:15 p.m. b 1:50 a.m., 1:35 p.m., 2:15 p.m. c 3:15 p.m., 4:45 p.m., 5:20 p.m. d 5:45 a.m., 6:40 a.m., 6:15 p.m. e 9:05 a.m., 8:55 p.m., 9:10 p.m. f 2:30 a.m., 2:50 a.m., 2:10 p.m. g 3:25 p.m., 3:30 p.m., 3:45 p.m. h 8:45 a.m., 8:55 a.m., 8:15 p.m. i 11 a.m., 11:10 p.m., 11:40 p.m. j 5:15 a.m., 5:25 a.m., 5:25 p.m.

3
a 12:45 p.m., 12:15 p.m., 10:15 a.m., 11:30 a.m.
b 11:30 a.m., 11 a.m., 9 a.m., 10:15 a.m.
c 2:15 p.m., 1:45 p.m., 11:45 a.m., 1 p.m.
d 1:30 p.m., 1 p.m., 11 a.m., 12:15 p.m.
e 2 p.m., 1:30 p.m., 11:30 a.m., 12:45 p.m.
f 12:15 p.m., 11:45 a.m., 9:45 a.m., 11 a.m.
g 1:50 p.m., 1:20 p.m., 11:20 a.m., 12:35 p.m.
h 12:35 a.m., 12:05 a.m., 10:05 p.m., 11:20 p.m.
i 12:10 a.m., 11:40 p.m., 9:40 p.m., 10:55 p.m.
j 2 a.m., 1:30 a.m., 11:30 p.m., 12:45 a.m.
k 1:45 a.m., 1:15 a.m., 11:15 p.m., 12:30 a.m.
l 1 p.m., 12:30 p.m., 10:30 a.m., 11:45 a.m.
m 12 midnight, 11:30 p.m., 9:30 p.m., 10:45 p.m.
n 11:45 p.m., 11:15 p.m., 9:15 p.m., 10:30 p.m.
o 2:10 a.m., 1:40 a.m., 11:40 p.m., 12:55 a.m.

Page 64

1 a 1735 b 2130 c 2340 d 0630 e 1045 f 1320 g 0815 h 0415 i 1410 j 1105 k 2255 l 1600 m 0750 n 1525 o 1000 p 0955

2 a 2:15 p.m. b 6:30 p.m. c 7:05 a.m. d 10:40 a.m. e 6:45 a.m. f 3:50 p.m. g 4:25 a.m. h 6:55 p.m. i 7:35 p.m. j 10:10 p.m. k 11:45 p.m. l 8:15 p.m. m 12:05 p.m. n 12:45 a.m. o 8:50 a.m. p 1:25 p.m. q 9:15 p.m. r 3:10 p.m. s 4:40 p.m. t 5:35 p.m. u 10:45 p.m. v 2:05 a.m. w 7:20 p.m. x 1:40 p.m.

3 a 0415 b 1750 c 2330 d 1315 e 0320 f 0745 g 1900 h 0935 i 2030 j 2300 k 2255 l 1810 m 1425 n 1645 o 0550 p 1000 q 1110 r 1535 s 0915 t 1305 u 1435 v 1600 w 1040 x 1210 y 0035

4 a 1 hour 20 min b 1 hour 40 min c 1 hour 15 min d 1 hour 50 min e 1 hour 50 min f 2 hours 10 min g 4 hours 30 min h 1 hour 35 min i 1 hour 50 min j 1 hour 50 min k 2 hours 15 min l 2 hours

5 a 1:20 p.m., 1:30 p.m., 2 p.m. b 0720, 0745, 0800 c 2005, 8:15 p.m., 2130 d 6:05 a.m., 6:15 a.m., 0640 e 1705, 5:15 p.m., 1755 f 2215, 10:20 p.m., 10:40 p.m. g 2030, 9 p.m., 2105 h 4 a.m., 1610, 4:30 p.m. i 3:30 a.m., 1540, 4 p.m. j 6:05 p.m., 1810, 1825 k 12:30 p.m., 1:15 p.m., 1345 l 1405, 2:30 p.m., 1450

Assessment

Pages 65–66

1 a 600 cm b 850 cm c 11.5 m d 40 mm
e 125 mm f 15 cm g $8\frac{1}{2}$ cm h 7000 m
i 10 500 m j 6 km k $19\frac{1}{2}$ km l $14\frac{1}{2}$ km

2 a 14 cm b 16 cm c 12 cm

3 a 14 m^2 b 17.5 m^2 c 13.5 m^2

4 a 5000 mL b 3500 mL c 750 mL d 8500 mL
e 6700 mL f 2150 mL g 1.5 L h 0.7 L
i 4.25 L j 4 m^3 k 9.5 m^3 l 7.25 m^3
m 100 000 cm^3 n 550 000 cm^3 o 675 000 cm^3

5 a 60 cm^3 b 54 cm^3 c 84 cm^3

6 a $3\frac{3}{4}$ kg, 3.5 kg, 3150 g, 350 g b 7.65 kg, 7500 g, $7\frac{1}{4}$ kg, 7.065 kg
c 8500 g, 1850 g, 1.8 kg, 0.85 kg d 5750 kg, 5.7 t, $5\frac{1}{2}$ t, 5000 kg
e 0.15 t, 115 kg, 0.1 t, 0.015 t f 2.8 t, 2080 kg, 2.008 t, 280 kg

7 a 570 seconds b 255 minutes c $2\frac{1}{2}$ hours d 152 hours
e 90 hours f 4 days g 56 days h 12 weeks
i 60 months j 39 months k 12 years l $7\frac{1}{2}$ years
m 35 years n 450 years

8 a $\frac{1}{2}$ past 9, 9:30 b 20 to 2, 1:40 c 25 past 1, 1:25

9 a 12:05 p.m. b 2:30 p.m. c 8:15 p.m. d 10:35 a.m.
e 6 p.m. f 7:20 a.m. g 12:10 a.m. h 3:40 p.m.
i 10:05 a.m. j 4:25 p.m. k 5:50 a.m. l 3:10 a.m.

10 a 11:50 a.m. b 9:35 a.m. c 1:45 p.m. d 8:20 a.m.
e 5:40 p.m. f 12:55 p.m. g 11:10 a.m. h 7:25 a.m.
i 5:50 a.m. j 4:05 p.m. k 11:35 a.m. l 8:55 p.m.

11 a 1:15 p.m. b 8:30 p.m. c 8:10 a.m. d 11:45 a.m.
e 3:25 p.m. f 10:35 p.m. g 5:05 a.m. h 12:45 a.m.
i 5:20 p.m. j 11:40 p.m. k 10:55 a.m. l 2:50 p.m.

12 a 0320 b 1445 c 2200 d 1150 e 1735 f 1930
g 0915 h 0610 i 1355 j 2105

Strand: Space and Shape

Pages 67–68

1 a triangle, 3, 3, 3 acute, 3 b square, 4, 4, 4 right, 4
c pentagon, 5, 5, 5 obtuse, 5 d hexagon, 6, 6, 6 obtuse, 6
e heptagon, 7, 7, 7 obtuse, 7 f octagon, 8, 8, 8 obtuse, 8
g nonagon, 9, 9, 9 obtuse, 9 h decagon, 10, 10, 10 obtuse, 10

2 a square prism, 8, 6, 12, No b rectangular prism, 8, 6, 12, No
c triangular prism, 6, 5, 9, No d cylinder, 0, 2, 0, Yes
e cone, 1, 1, 0, Yes f cube, 8, 6, 12, No
g square pyramid, 5, 5, 8, No h sphere, 0, 0, 0, Yes
i triangular pyramid, 4, 4, 6, No

Page 69

1 b **2** a **3** a and b **4** b and c
5 a and b **6** a

Page 70

1 a, e and m are equilateral triangles

2 b, f, h, j, o and q are isosceles triangles

3 c, d, g, i, k, l, n, p, r and s are scalene triangles

4 c, f, l, n and s are right-angled triangles

Page 71

1 c, d, e, f, h, j, k, l, m and n have horizontal symmetry

2 a, b, c, d, e, g, i, j, k, l, o and p have vertical symmetry

3 c, d, e, j, k and l have both

Page 72

1 a acute b obtuse c reflex d right e straight f right
g obtuse h obtuse i acute j reflex k straight l acute

2 a obtuse b acute c right d reflex e acute f acute
g right h acute i obtuse j obtuse k acute l obtuse

Page 73

Estimates will vary.

1 45° **2** 90° **3** 120° **4** 60° **5** 80°
6 25° **7** 100° **8** 30° **9** 50° **10** 90°
11 70° **12** 40° **13** 130° **14** 10° **15** 75°
16 140° **17** 160° **18** 85° **19** 95° **20** 15°

Page 74

1 **2** **3**

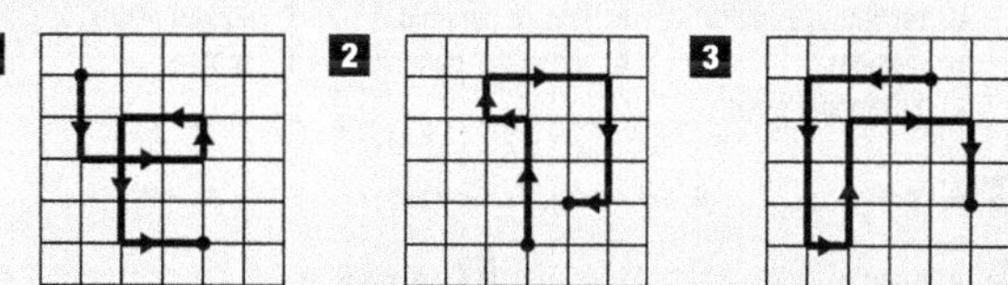

4 **5** **6**

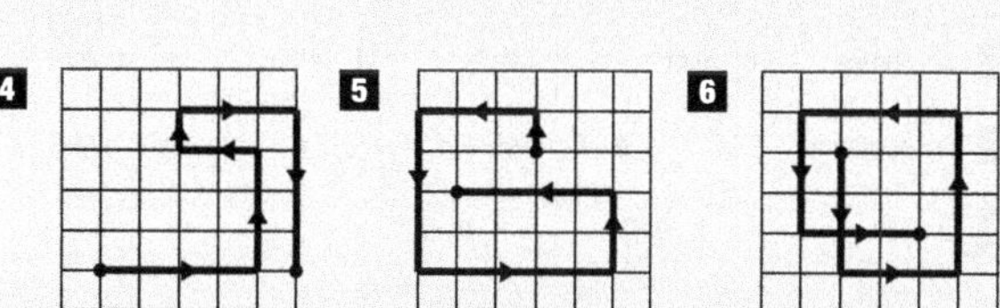

7 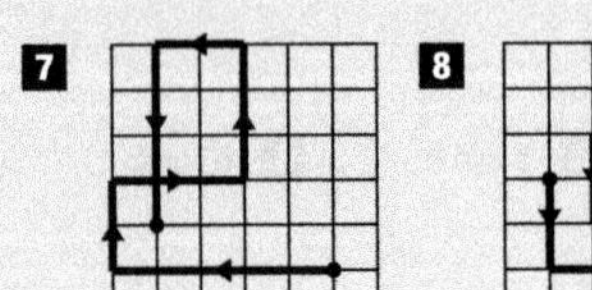 8 9

9 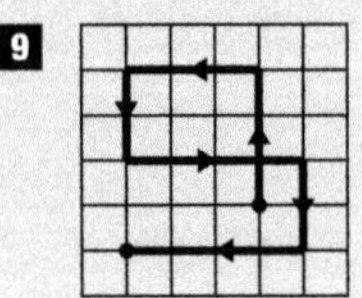

10 2 north, 3 east, 2 north, 2 west, 3 south, 3 east

Page 75

1 2 north, 3 east, 2 south, 1 east, 4 north, 3 west

2 4 east, 2 south, 3 west, 2 south, 3 east, 1 north

3 4 north, 2 east, 3 south, 4 west, 2 north, 1 west

4 5 north, 3 west, 4 south, 2 west, 3 north, 3 east

5 2 west, 3 south, 4 east, 2 north, 1 west, 3 north

6 5 west, 3 south, 2 east, 4 north, 1 east, 3 south

7 2 south, 2 east, 4 north, 2 east, 2 south, 1 west

8 4 west, 3 north, 2 east, 4 south, 1 east, 5 north

9 3 north, 3 east, 1 south, 4 west, 3 south, 4 east

10 2 south, 4 west, 1 north, 3 east, 4 south, 2 west

11 2 east, 4 south, 4 west, 4 north, 2 west, 3 south

12 3 south, 4 east, 4 north, 3 west, 1 north, 2 east

Assessment

Pages 76–77

1 **a** pentagon **b** octagon **c** rectangle **d** circle **e** hexagon

2 **a** triangular prism **b** cube **c** cone
d rectangular prism **e** square pyramid **f** square prism
g cylinder **h** triangular pyramid **i** sphere
j hexagonal prism

3 **a** yes **b** no **c** yes **d** no

4 a, c and e

5 a, b and d

6 **a** obtuse **b** right **c** acute **d** reflex **e** straight

7 **a** heptagon **b** 7 angles **c** 3 are acute angles

8 **a** 45° **b** 120° **c** 90°

Strand: Chance and Data

Page 78

1 **a** 12 **b** 23 **c** 17 **d** 34 **e** 27 **f** 36

2 **a** Elsie = 16.6; Peter = 13.1; Ben = 18.3; Imata = 15.7; Philip = 19.1; Taita = 14.6; Regina = 16.1; Nancy = 18.8
b Philip **c** Peter

3 **a** Port Moresby = $31\frac{5}{7}$; Rabaul = $30\frac{6}{7}$; Lae = $28\frac{1}{7}$; Goroka = $25\frac{4}{7}$; Mt. Hagen = $22\frac{1}{7}$
b Port Moresby

Page 79

1 **a** 17 **b** 27 **c** 51 **d** 13 **e** 40
f 61 **g** 71 **h** 46 **i** 29 **j** 89

2 **a** 3 **b** 12 **c** 18 **d** 27 **e** 33
f 81 **g** 2.5 **h** 6.1 **i** 5.7 **j** 2.6

3 **a** 8.5 **b** 17 **c** 30 **d** 31 **e** 79.5
f 49 **g** 69.5 **h** 89.5

4 **a**

Shoe size	Number sold
5	2
$5\frac{1}{2}$	2
6	3
$6\frac{1}{2}$	5
7	5
$7\frac{1}{2}$	9
8	9
$8\frac{1}{2}$	3
9	4
$9\frac{1}{2}$	1
10	2

b $7\frac{1}{2}$ and 8 **c** $7\frac{1}{2}$ and 8

Strand: Patterns

Page 80

1 a

5	8	3	12	17	9
10	13	8	17	22	14

Rule = add 5

b

20	11	18	23	15	19
13	4	11	16	8	12

Rule = subtract 7

c

6	10	7	13	2	25
60	100	70	130	20	250

Rule = multiply by 10

d

25	40	15	30	50	100
5	8	3	6	10	20

Rule = divide by 5

e

4	6	3	11	7	20
16	36	9	121	49	400

Rule = square each number

f

16	20	8	36	80	28
8	10	4	18	40	14

Rule = halve each number

g

7	12	25	9	14	31
21	36	75	27	42	93

Rule = multiply by 3

h

340	230	510	290	460	370
34	23	51	29	46	37

Rule = divide by 10

i

9	15	34	40	28	19
18	30	68	80	56	38

Rule = double

j

17	14	21	38	26	47
67	64	71	88	76	97

Rule = add 50

2 a

4	6	10	5	2	9
10	14	22	12	6	20

Rule = double and add 2

b

3	5	2	10	6	8
10	16	7	31	19	25

Rule = multiply by 3 and add 1

c

8	12	15	9	20	25
15	23	29	17	39	49

Rule = double and subtract 1

d

9	27	18	30	15	36
4	10	7	11	6	13

Rule = divide by 3 and add 1

e

16	20	8	24	14	10
7	9	3	11	6	4

Rule = halve and subtract 1

f

60	30	100	70	80	120
7	4	11	8	9	13

Rule = divide by 10 and add 1

g

5	10	8	20	12	15
13	23	19	43	27	33

Rule = double and add 3

h

30	100	18	60	70	40
17	52	11	32	37	22

Rule = halve and add 2

i

60	48	100	26	18	40
27	21	47	10	6	17

Rule = halve and subtract 3

j

6	9	3	7	11	2
65	95	35	75	115	25

Rule = multiply by 10 and add 5

Page 81

1 a 45, 50, 55 b 56, 46, 36 c 35, 41, 48
d 17, 19, 20 e 35, 25, 20 f 81, 243, 729
g 82, 64, 128, 256 h 48, 96, 192, 384 i 24, 23, 21, 20
j 63, 62, 66 k 175, 351, 703 l 65, 129, 257
m 24, 27, 29 n 54, 53, 48 o 72, 216, 432, 1296
p 125, 253, 509 q 59, 60, 50, 51 r 37, 36, 46
s 3.2, 6.4, 12.8, 25.6 t 4.8, 9.6, 19.2

2 a add 9 b add 11 c multiply by 4 then multiply by 2
d subtract 6 e subtract 2 then subtract 3
f add 5 then add 10 g double and add 1
h double and subtract 1 i multiply by 3 and add 1
j subtract 10 then subtract 5 k add 1 then add 2 then add 3…
l subtract 1 then subtract 2 then subtract 3 …
m add 9 then subtract 1 n subtract 9 then add 1
o double p triple
q add 1 then add 0.5 r subtract 0.2 then subtract 0.5
s add 2 then subtract 0.5 t double then add 0.5

3 a – t Teacher to check.

Page 82

1 a 36 b 20 c 52 d 173 e 8 f 2
g 5 h 54 i 30 j 60 k 14 l 16
m 16 n 36 o 46 p 28 q 20 r 5
s 8 t 30 u 3 v 7 w 12 x 6

2 a $5 \times (6 + 4) = 50$ b $40 - (6 + 4) = 30$ c $(10 + 2) \times 5 = 60$
d $(10 + 15) \div 5 = 5$ e $(12 - 2) \times 4 = 40$ f $10 - (5 + 3) = 2$
g $8 \times (4 + 6) = 80$ h $5 \times (8 + 4) - 5 = 55$ i $36 - (6 + 10) = 20$
j $24 \div (6 + 2) = 3$ k $(6 + 8) - (3 + 9) = 2$ l $(20 - 4) \times 3 = 48$
m $\frac{1}{2}$ of $(8 + 2) = 5$ n $6 \times (10 - 8) = 12$ o $(7 + 3) \times 4 = 40$
p $5 \times (8 + 2) = 50$ q $(20 - 8) \times 4 = 48$ r $(30 - 6) \div 2 = 12$
s $5 \times (3 + 3) - 2 = 28$ t $(7 - 3) \times (7 + 2) = 36$ u $20 \div (2 + 2) = 5$
v $(3 + 5) \times (5 - 1) = 32$ w $7 \times (4 + 4) - 6 = 50$ x $(18 - 9) \times 10 = 90$

3 a = b ≠ c = d = e ≠ f =
g = h ≠ i = j = k ≠ l ≠
m ≠ n = o = p = q ≠ r =
s ≠ t = u = v = w ≠ x =

4 a 10 b 59 c 44 d 63 e 44 f 11

Assessment

Page 83

1 b **2** c **3** b **4** a **5** d **6** c
7 b **8** c **9** a **10** a **11** d **12** c
13 b **14** c **15** b **16** d **17** a **18** c
19 d **20** a